A Treatise on Ecological Science

Authored By

Vikas Rai

Department of Mathematics
Eritrea Institute of Technology
Mai - Nefhi, Eritrea

A Treatise on Ecological Science

Author: Vikas Rai

ISBN (Online): 978-981-5322-49-1

ISBN (Print): 978-981-5322-50-7

ISBN (Paperback): 978-981-5322-51-4

© 2024, Bentham Books imprint.

Published by Bentham Science Publishers Pte. Ltd. Singapore. All Rights Reserved.

First published in 2024.

need for a court order if at any point you breach any terms of this License Agreement. In no event will any delay or failure by Bentham Science Publishers in enforcing your compliance with this License Agreement constitute a waiver of any of its rights.

3. You acknowledge that you have read this License Agreement, and agree to be bound by its terms and conditions. To the extent that any other terms and conditions presented on any website of Bentham Science Publishers conflict with, or are inconsistent with, the terms and conditions set out in this License Agreement, you acknowledge that the terms and conditions set out in this License Agreement shall prevail.

Bentham Science Publishers Pte. Ltd.
80 Robinson Road #02-00
Singapore 068898
Singapore
Email: subscriptions@benthamscience.net

BENTHAM SCIENCE

CONTENTS

FOREWORD

When we started to plan our course "Dynamic Models in Biology", and the textbook of the same name to go with it (Princeton University Press, 2006), my co-author and co-instructor John Guckenheimer observed that "the best books are often the most personal books". That is: books aiming for even-handed, comprehensive coverage of a subject are useful as reference books -- I have a few such on my bookshelf, many with "encyclopedia" or "handbook" in their titles, and some of them really have been very useful. But books that very selectively highlight one person's idea of what is important in a subject area, one person's interpretation of current knowledge, and their projections about the future – those are the books that get read from cover to cover, which can divert a reader into a different line of inquiry, and even change the direction of an entire field.

Vikas Rai has written a very personal book about ecology. The viewpoint is modern. Classically an ecosystem was typically conceptualized as stocks and flows of various molecules (C, N, P, *etc*.). That view is not without value when our goal is to summarize global biogeochemistry. But we now know how crucial it is that those molecules are clustered into individual organisms, who attend to the world around them and make intelligent decisions, shaped by natural selection. The fundamental events in ecosystems are then births and deaths, that is: consuming, being consumed, and converting consumed resources into offspring. Accordingly, the book opens with the study of these interactions and the competition that results when there are multiple consumers for one food item. The subsequent chapters broaden from this fine scale to networks of interacting species, and then to the level of global biodiversity. Individual actions – in particular human individual actions – again take center stage in the next three chapters, exploring how human decisions now determine the state and fate of planet Earth. The last chapter concerns sustainable development goals; the agenda set by Comity of Nations; UNO.

So will this book teach you about all of ecological science? No, not even close. But it will tell you how one person views those topics after decades of thinking about them and contributing to them, and what he thinks is most important. And maybe it will help you choose how you want to make your own contributions.

Stephen P. Ellner
Professor of Ecology and Evolutionary Biology
Cornell University
NY 14850, USA

PREFACE

A textbook on ecology that covers all aspects of ecological sciences has been missing in the literature. It will be befitting to bring out such a text that prepares for an introductory class of undergraduate students. For this, the knowledge of matrix theory and differential equations (both ordinary and partial) is the prerequisite. The present book lacks a chapter on 'microbial ecology'. The author has no expertise in the subject. Details of microbes and their interactions with plants can be found in other texts. A concise book with essential details and lucid descriptions of the latest knowledge on key topics is bought out.

Ecological Imbalance is the principal cause of most of the problems on Earth. Ecosystem services are impeded in several known and unknown ways when such an imbalance occurs. In order to understand how this causative agent operates, key elements of 'ecological complexity' are required to be understood. The present book provides a framework to understand the 'balance of nature' with minimal use of mathematics in order to reach a broader readership.

Ecological systems are an example of 'complex systems'. The content of the chapter presents **ecosystems** as the 'network of networks'. The chapter on network ecology provides a brief description of how interaction types and the number of interactions among species determine its 'existential capacity' and 'functional efficiency'. The key elements of an ecosystem are food webs; a network of interconnected food chains. The present text classifies food chains into two types: *linear and nonlinear.*

Climate change drives catastrophic changes in biodiversity. This is the reason why the book presents a brief chapter on biodiversity and its relationship with **climate change**in Chapter 2. The third chapter on human ecology provides a framework to integrate *ecological sciences with neurosciences*. Modern civilization cannot exist without industry. Industries cause pollution which deteriorates the **'quality of life'** on Earth. The author emphasizes that it must be taught at the undergraduate level so that the basic philosophy of the subject is injected into the DNA of individuals of *homo - sapiens*. This is the fourth chapter.

The last chapter on Sustainable Development Goals (SDG) provides a historical perspective on the topic of 'sustainable development'. Under SDG 1 (**No poverty**), both the definitions of poverty; workable and broader, are provided. Ecological viewpoints of all SDGs are discussed. In sum, after reading all the chapters in the book, a student will have sufficient knowledge to analyze the phenomenal world around him/her.

Vikas Rai
Department of Mathematics
Eritrea Institute of Technology
Mai - Nefhi, Eritrea

CHAPTER 1

Network Ecology

Abstract: Ecological systems (**populations and communities**) interact with each other. These entities can be viewed as networks and ecosystems as 'networks of networks'. Ecological networks share common properties with other networks; *e.g.*, Wireless Sensor Networks (WSNs). WSNs consist of receivers and transmitters of information at locations called **nodes.** These nodes transmit and receive information with each other in 'packets'. In the context of ecology, these packets contain material and energy; *e.g.*, the bird from the bird sanctuary (20 kms away from my residence) being caught by the cat for food. Elements of network theory which are essential for applications to ecological networks are introduced. Decisions of animal movements and observed patterns of movement can be better explored in this framework. Although these networks have complex architecture, their hierarchical nature admits well-defined patterns that illuminate mechanisms of functioning of **ecosystems**. Applications of network theory would advance the understanding of complex interactions between species; 'tangled banks' of nature.

Ecological networks are simulated. These simulation experiments illuminate observed patterns of movement. *A network of social interactions* and a *network of movement patterns* are explored to know how movement decisions are taken.

Keywords: Biodiversity, Energy, Layering, Food chains, Food webs, Host-Parasite interactions, Interaction Types, Link distribution, Loss of resilience, Levels of organization, Material, Monolayer network, Multilayer, Multilevel, Networks, Resilient, Resilience, Strong trophic interactions, Temporal food – webs, Weak – trophic interactions.

INTRODUCTION

Investigations in ecological theory and modeling have focused largely on simple food – webs; two or more food - chains linked with migration or with a few *weak – trophic interactions*. Natural ecological systems are more complex than one can imagine. Ecological systems consist of several weak (predation/harvesting) and a few *strong trophic interactions* (McKann, 1998; O'Gorman & Emmerson, 2009). Network ecology presents a framework for conceptualization and analysis of ecosystem function and its response to external perturbations. The concept of ecosystem refers to both the flow of energy and to species interactions. Interaction strengths in complex food webs are discussed by Berlow *et al.* (2009). Before we begin to describe different types of networks, definitions of essential elements of network theory are described below.

Graph Pair-wise relations between two objects.

Degree Degree of the vertex of a **graph** is the number of edges incident on it.

Edge A link between two vertices in a network. When weighted denotes interaction strength.

Vertices A point where two or more lines, curves, or edges meet.

Flow dynamics The movement of resources (energy and information) on a network.

MONOLAYER NETWORK

Farage, *et al.* (2021) have presented a scheme to identify flow modules in ecological networks using *infomap*; a software package for landscape analysis. Fortin *et al.* (2021) have discussed the dynamics of ecological networks in response to environmental changes. It focuses on how the interplay between species interaction networks and spatial structures of landscape patches explains the observed variability in space and time. Newman and Girvan (2004) proposed algorithms to divide network nodes into densely connected subgroups, which represent community structure. The authors presented a measure to determine the strength of identified community structures. Readers are referred to a book on the subject by Newman (Newman, 2010). Algorithms have been designed to remove edges from the network iteratively. These algorithms are based on graph partitioning ideas. It finds groups of nodes with dense connections with sparser intergroup connections. The network is broken into its communities by these clustering algorithms. Newman and Girvan also proposed a measure to determine the strength of communities thus identified. A typical monolayer network is shown in Fig. (**1**).

A food - web consists of food – chains linked with migration (Kitching, *et al.* 1987). Food chains are of two types: ***linear and non-linear***. Fig. (**2**) shows a representative terrestrial food - web which exemplifies a monolayer network given in Fig. (**1a**). It consists of three food chains (Murray & de Ruiter, 1991).

1. Grass - Grasshopper – Frog – Hawk
2. Grass – Mouse – Hawk
3. Grass – Rabbit - Hawk

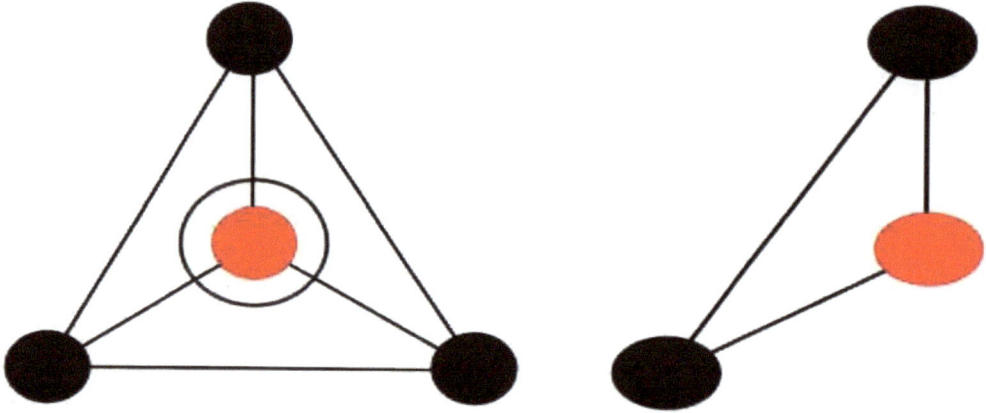

Fig. (1). (a) Connected nodes. Nodes represent patches. Species dispersal in a patchy environment is shown. (b) A Monolayer Network. The keystone species patch is shown in red. A keystone species is a critical source of food for predators; *e.g.*, Antarctic krill, Canadian snowshoe hares, *etc*. Wolves and sea otters are predators that control populations and the range of their prey. Loss of a keystone species means loss of ecosystem services as it would no longer be functional. Three monolayer networks are linked together.

Hawk is a keystone species. Grass is the primary producer. Grasshoppers and frogs are herbivores. Hawk is a predator. Grass – Grasshopper system is the consumer resource system. Frog–Hawk is an example of a weak trophic interaction. Fig. (**2**) presents a food chain with these species. Fig. (**3**) presents an aquatic food chain.

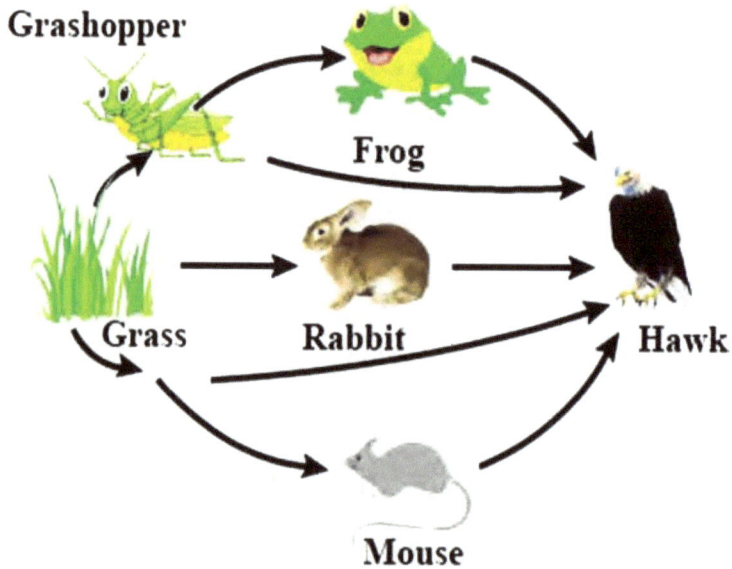

Fig. (2). A Terrestrial Food – web (part of an ecosystem). It consists of many food chains.

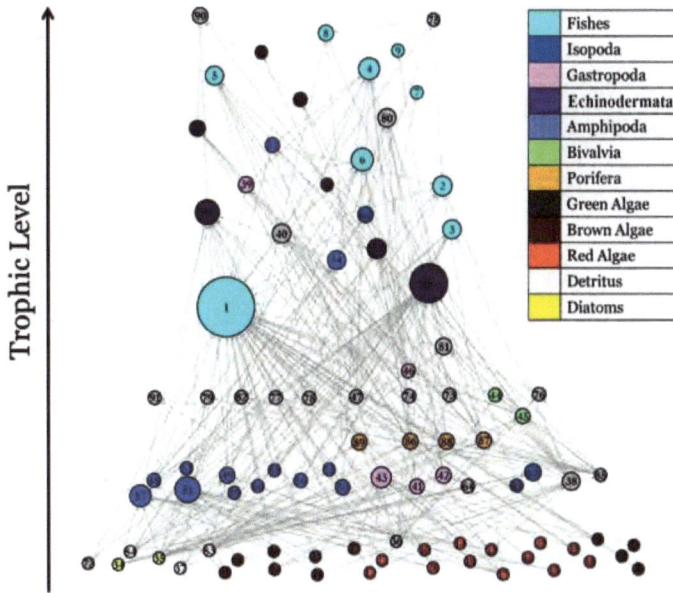

Fig. (3). Network structure of an Antarctic Food – web. Fish is the '**keystone species**' (Source: Wikimedia).

TYPES OF FOOD CHAINS

Linear food chain. Oscillation at a trophic level induces the same type of oscillation at the higher trophic level.

Nonlinear food chain. Oscillation at a trophic level does not necessarily induce the same type of oscillation. Induced oscillations could be nonlinear oscillations (a stable limit cycle) or chaotic oscillations, which consist of many unstable oscillations generating order. Breaking down a food web structure into linear and non-linear food chains holds the key to the analysis of both these properties: *EngineeringandEcological resilience*. A few key terms are introduced to the reader before he/ she attempts to apply network theory to ecological modeling.

Coupled dynamics. It represents a change in the topology of the network and the flow of resources through it. It also includes feedback mechanisms between these two processes operating on the same time scale. The frequency distribution of degrees can sometimes indicate underlying dynamic processes.

MULTILAYER NETWORK

It is a network with several types of connections between nodes. Different layers of the network represent different interaction types. In case there is only one layering aspect in a multilayer network, and nodes are coupled diagonally, this type of network is called multiplex (cf. Fig. **4**). A multilayer network is a quadruplet, $M=(V_M, E_M, V, L)$, where, $\mathbf{L} = \{L_a\}$; a runs from 1, is a sequence of sets L_a of elementary layers. a is an index for aspects of layering. An elementary layer represents a single aspect in any given layer. V is the set of physical nodes. V_M is a set of state nodes. It encodes manifestations of an entity $v \epsilon V$ on a particular layer, l. The edge set E_M encodes connections between pairs of state nodes. The intra-layer edges encode connections of a particular type.

SEVERAL ASPECTS OF LAYERING

Layers could be defined either in space or in time. Adaptation and application of network theory to ecological networks depend on our ability to distinguish between intra-layer and interlayer connectivity. Ecological systems have two intrinsic properties: high – dimensionality and heterogeneity. Layers defined by interaction types define a community's structure. An example of a multi-layer network is shown in Fig. (**5**).

A brief description of ecological interactions is provided as layers, which are defined by interaction types.

Symbiosis (Mutualistic)

Plant – pollinator

Fish and anemone

A variety of interactions occur together. Mice build nests. Bumble bees use them. This is an example of ecological engineering. An increase in seed set in Clover is a **mutualistic pollination interaction** (Bastolla *et al*, 2009; Holland *et al*., 2002; Trojelsgaard *et al*., 2015).

Predation

Cats on mice

Fox – rabbit

Predator-prey communities; foxes preying on rabbits, and voles preying on weasels in the boreal Fennoscandanavia (Ylonen, *et al*., 2019) are a few examples.

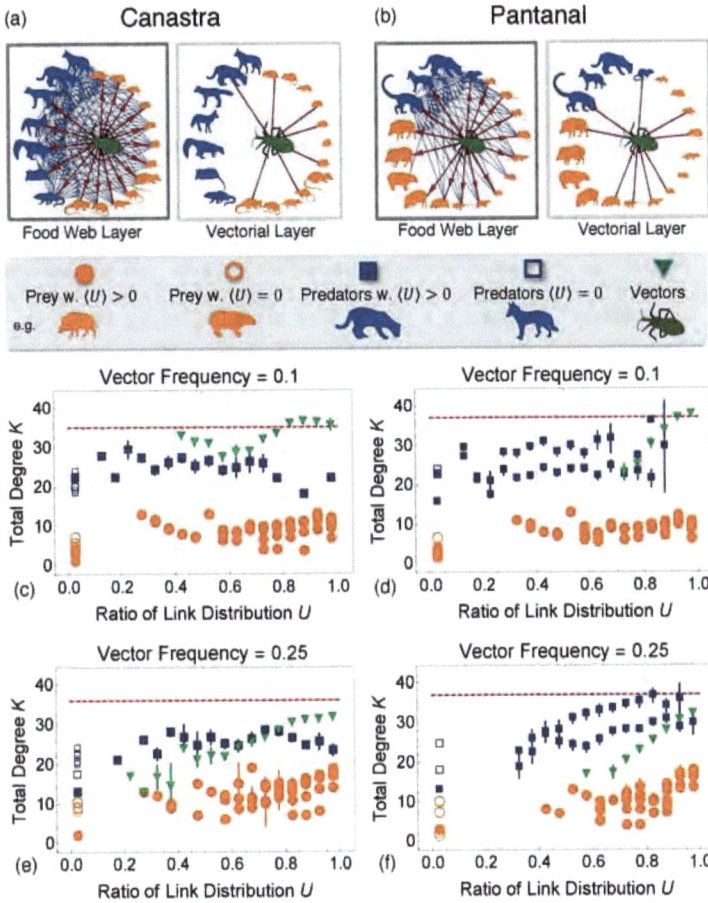

Fig. (4). Multiplex networks. Canastra is a type of grass which is found in Brazil. Pantanal is a region which houses the world's largest wetland area. It is spread over Brazil, Bolivia, and Paraguay.

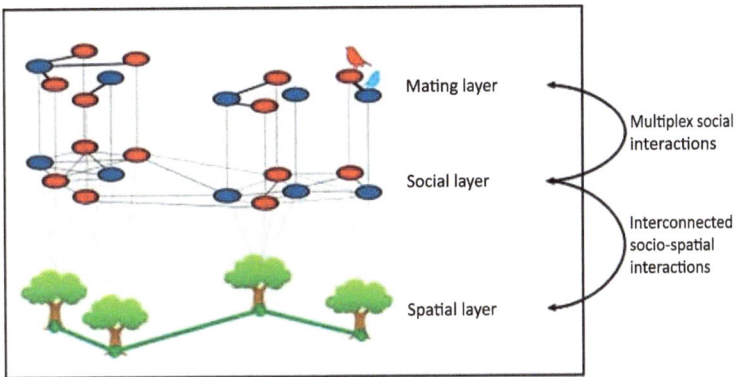

Fig. (5). A multi-layer Network (Courtesy: Cell Press; Silk *et al.*, *Trends in Ecology and Evolution*, vol. 33, no. 6, pp. 376 – 378, June 2018).

Parasitism

E. O. Wilson (2014) defined parasites as predators that eat prey in units of less than one. The interaction manifests itself in such a way that an organism lives either on the body of its host or inside it. There is rarely an actual death. Single-celled protozoa; agents of malaria, honey fungus, hookworms, lice, and mosquitoes; plants such as mistletoe, dodder, and broomrapes, *etc*. are a few examples. **Host-parasite interactions** (Sait, *et al*., 2003) promote disease tolerance to intestinal helminth infection (King and Li, 2018). Host – Parasitoid foraging and reproduction are directly linked with each other (Cook and Hubbard, 1977; Hubbard and Cook, 1978). Luck *et al*. (1999) present a detailed description of these interactions.

Competition

Competition between and among species are described by Gause type of models. Intra and interspecific competition is common in certain habitats; *e.g*., drylands. Water is the limiting resource. It is the limiting resource in Keoladev National Park, Bharatpur, India. Van - dermeer (2002) pointed out that increased competition might promote species coexistence. Local exclusion of species results from large competitive pressure. The orthodoxy of linear thinking is dispelled when nonlinear thinking is invoked. A high level of competition may actually prompt species to adopt behaviors that facilitate avoidance of local exclusion.

Non-trophic interaction is an interaction that does not involve the transfer of energy between two individuals participating in the interaction.

A brief description of a few types of ecosystems is given below.

GLOBAL DRYLANDS

Vegetation occurs in patches. Competition between plant species for water, the limiting resource, is high. Herbivores consume these plants. Cats influencing seed sets in Red Clover is an example of an indirect effect (Ulrich *et al*., 2019).

Plants Three–leaved cinquefoil, *Danthonia spicata*, Crowberry, Deergrass, Cranberry, grasslands, empetraceae, *etc*.

Animals lizards and invertebrates

Birds Banded dotterels, pipit, black-fronted terns (Bonkoungou)

Savana

Woodland – The Grassland ecosystem is characterized by a wide spacing of trees, where canopies do not close. It facilitates the growth of grasses as sunlight reaches the ground. Elephants, Lion, wildebeests, and several mammal species are found.

Endangered Species Diceros, African bush elephant, white–backed vulture, African wild dog, southern white rhinoceros, *etc.* are a few endangered species (Walker, *et al.*, 1981).

Climate

Climate is temperate in mid-latitudes, between the Tropics and the Polar regions of the world. It is neither extremely hot nor cold. Temperate regions (40-degree to 70-degree latitudes) witness moderate rainfall spread across the year. **Plant Species** include *Lyonia Ovalifolia*, European beech, Sweetgum, filbert, Strawberry tree, Quaking aspen, English oak, *etc.*

Animal Species. Ring–billed gull (a bird), swamp rabbit, and nutria are found in the scrub near the beach.

Coastal Ecosystems. *Oceans* maintain the temperature of the land environment.

Forest Ecosystems. Owls, bats, and raccoons are a few dominant species in the night.

MULTILEVEL NETWORKS

Diversity at three levels

1. Species diversity,
2. Genetic diversity, and
3. biological diversity

Inter-layer edges represent species dispersal between communities. A species can also disperse to a new community.

Ecological Complexity

1. Multiple interaction types,
2. Interactions that vary in space and time, and
3. Interconnected systems; *e.g.*, network of networks

NETWORK OF NETWORKS

Communities in different habitats and Patches are linked with each other. The network structure of an individual patch is developed. Networks corresponding to two patches are connected with each other and are extended in space and time. The movement of animals in patchy environments (***movement patterns and movement decisions***) can be explored (Ke'fi *et al*. 2012, 2017) using a network of networks. An idiosyncratic feature of these is **complexity in space and time**; *e.g*., flowers in one season and the other are visited by different pollinators, and the same plant species may interact with both pollinators and herbivores. Algorithms for the detection of community structure in networks with edges representing different types of interaction are available in the literature (Newman & Girvan, 2004; Newman, 2010).

NETWORK ECOLOGY

Meta-communities are a set of interacting communities that are linked by dispersal. Multiple interaction types which govern the dynamics of interacting populations vary both in space and time. The dynamics of interacting populations are modeled as ***networks*** with a single interaction type.

ARCHITECTURE AND DYNAMICS OF ECOLOGICAL SYSTEMS

Layers are defined by interaction types (Mougi, 2012). Competition for common resources determines the number of coexisting species. A Consumer-resource model with facilitative interactions determines the total system biomass and thus the way ecosystems function (Gross, 2008, Ke'fi, *et al*. 2012). Systems robustness to extinctions is enhanced if edge effects signifying ontogenetic shifts are also included (Rudolf & Lafferty, 2011). Do food webs have characteristic path lengths? Clustering coefficients and distribution of degrees are parameters to qualify to be a member of this class. Dunne *et al*. (2000) analyzed data from food webs with 25-172 nodes. Terrestrial and aquatic ecosystems have low clustering. Food webs with ratios of observed to random clustering coefficients increase as a power law with network size over 7 orders of magnitude. Chamacho, *et al*. (2002) suggested that scaling and universality may be good descriptors of ecosystems. Universality classes exist for ecosystems from different environments.

COMPLEX ECOLOGICAL NETWORKS

1. Networks of single interaction types
2. Multiple interaction types to understand community structure and dynamics. Plants compete for water, *the limiting resource*. Diverse interaction types link species in natural habitats (Ke'fi, *et al*., 2017).

Early work explored how the composition of species changes with time (*e.g.*, across seasons or over environmental gradients). Later studies have explored spatial and temporal dissimilarity in species and interactions. Biological processes at any given level of organization (*e.g.*, genes, individuals, populations, *etc.*) can depend on processes at other levels (O'Neil, 1986). For example, changes in species biomass can affect the stability of food webs in dynamical models based on allometry (Brose, 2010).

In case, layers in a multi-layer network represent different levels of an organization, a 'multi-level' network interactions among nodes at lower levels automatically entail interactions at upper levels (Kivela¨, 2014). A trophic interaction between two species from two different patches implies that there is an interaction between patches; *e.g.*, a two-level network, can also be constructed as a network of networks (Poisot *et al.*, 2012). A multi-level network is shown in Fig. (**6**).

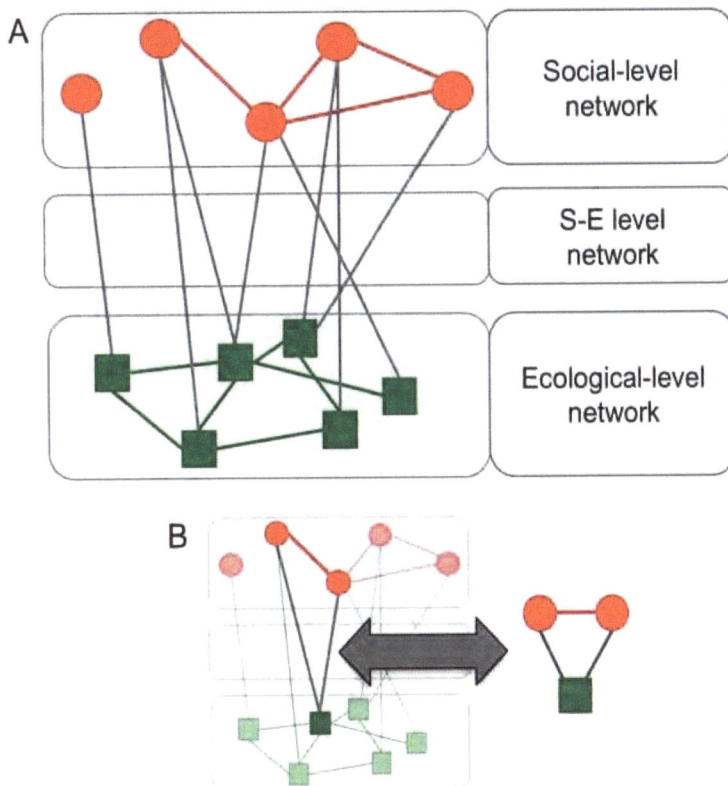

Fig. (**6**). **A multi - level network.** The multi-level structure displays dependencies within the system. Constraints and opportunities presented by the system structure for social action are presented.

METHODOLOGY

Layers are defined by interaction types and by levels of organization. The interconnectedness and architecture of ecosystems are visible in Fig. (**6**). The flow of materials is not shown, still one gets an idea of in what directions it would take place. Simulations of models describing community interactions lead to a better understanding of the dialogue between ecological and social networks. An ecological network can be explored in R. The number of 'edges' incident on a **node** is regarded as its degree (Blonder *et al*, 2012) in this system. The efficacy of biological control measures can be estimated if sufficient experimental data is available (Luck *et al.*, 1999). The following can be studied:

1. The effect of intra-layer and or interlayer connectivity on community structure,
2. The modular structure of the community in (1) on results of random associations between hosts and parasites in any given layer, and
3. Independence of the temporal order in which hosts and parasites are observed.

Rejection of these hypotheses helps in the understanding of the functional groupings of species by demonstrating that host–parasite interactions (Govaert *et al.*, 2021; Starr & Cline, 2002) are structured *non-random* in time and depend on how species interact in a given time period and their persistence in time (measured by changes in species abundance). Govaert *et al.* (2021) focused on the eco-evolutionary dynamics in Freshwater Systems.

ROBUSTNESS TO PERTURBATIONS

O'Gorman and Emmerson (2009) have explored how perturbations to trophic interactions drive changes in the stability of complex food-webs. To test whether the stability of ecological communities is affected by network structure, a network of two layers; plant – flower – visitors and plant – leaf – miner parasitoids that are interconnected with the same set of plants, is considered. Patterns of ***parasitoid extinctions*** are explored in two scenarios (Pilosof, *et al.*, 2017)

1. Direct secondary extinctions due to plant removal (a monolayer scenario) and
2. Tertiary extinctions due to the removal of pollinators, which cause plant secondary extinctions.

Disconnections from flower visitors lead to extinction of plants and parasitoids. These extinctions are found to be slower in the multi-layer network in comparison to when plant - the parasitoid layer is examined as a monolayer network. Non-trivial extinction patterns are obtained when plant extinctions are considered with those of flower visitors in the latter scenario. The interplay between different interaction types and their effect on the system's robustness can be explored.

BIOENERGETICS OF TEMPORAL FOODWEBS

Food-web structure and function varies with time. Several investigators have explored relationships between the structure and dynamics of food – webs (Winemiller & Polis, 1996, Winemiller & Jepsen, 1998). Comparative studies between the temporal dynamics of food webs in different ecosystems were performed (Kortsch, 2021). Biochemical contents of prey species were estimated in the Patagonian fjord food web (Ruiz - Ruiz, 2021).

Temporal Food-Webs

Using long–term biomass data pertaining to trophic levels of more than three, with sufficient information on species feeding and foraging behavior, Kortsch *et al*. (2021) analyzed food–web dynamics in the Gulf of Riga (Baltic seas) over the period (1981 – 2014). Temporal food web dynamics were explored by these investigators using unweighted (topology-based) approach and weighted (biomass and flux-based) approach. These authors present a lucid description of how temporal dynamics is governed by food-web structure. **Ecological Network Analysis** is performed after integration of unweighted, node – and link – weighted metrics to compare food – web structure and function. Patterns of change in food web dynamics are captured.

Biochemical contents and energy flow in a food web depend on the physiological characteristics and feeding strategies of its species. Biochemical composition of prey is an important factor since it serves as food for species at higher trophic levels. Ruiz–Ruiz (2021) studied traits of three keystone marine species in the Patagonian fjord food web. It was found that *M. gregaria* has higher concentrations of lipids, proteins, glucose, and total energy compared to either *E. maclovinus* or *C. australis*. Availability of food, environmental conditions, and physiological characteristics are key factors that make the presence of lipids in all the species possible.

Uncertainty about habitat carrying capacity, the proliferation of chemicals and contaminants and the emergence of hybrid food webs containing native and invasive species are three main factors that impede successful river restoration. An understanding of the causes and consequences of changes in nutrients, organic matter (energy), water, and thermal sources and flows is instrumental in working out a scheme of restoration efforts (Naiman *et al*., 2012).

RESILIENT NETWORKS

These are connected resilient systems. In what follows, these systems are described briefly.

Resilient System

A resilient system is one which returns to its original state after a perturbation is applied to it. The perturbation could be large enough to cause catastrophic effects in an ecological system (cf. Scheffer *et al.*, 2001). The path of return to the original state could be cyclic or hysteresis. It would be interesting to examine these possibilities as the conservation of biodiversity is a priority. The amplitude of consumer – resource oscillations is controlled by weak trophic interactions. Population extinctions are avoided by maintaining population densities away from zero (McKann, *et al.* 1998, McKann, 2007). Food-web interaction strengths are characterized by many weak interactions and a few strong interactions. The existence of a weak interaction for every strong consumer–resource interaction holds the key for intact ecosystems.

The time taken by the system to return to its original state after it is acted upon by an external perturbation is a key observable. There could be multiple routes for the system to return to its original state. C. S. Holling (1996) developed concepts of Engineering Resilience and Ecological Resilience in order to understand responses of functional ecosystems to external perturbations. Gunderson (2000) contrasted these two concepts so that others could understand it well. Zampieri's attempt (Zampieri, 2021) to reconcile these two concepts is directed towards putting them to practice. Concepts and tools of network analysis have been applied to biological control of weeds. Reliable and resolved ecological networks are constructed with the help of suitable methods.

Resilient Ecological networks could be characterized by engineering resilience (Gunderson, 2000; Holling, 1996; Rai, 2013; Zampieri, 2021). Stability near an equilibrium steady state is studied empirically to assess the resistance of an ecosystem to external perturbations (Walker *et al.*, 1969). Ecosystems possess multiple steady states. The organization of these states in the phase space of an ecological system decides its response to external perturbations. The resilience of an ecological system was discussed in detail by Rai (Rai, 2013). A complex set of weighted; directed interactions define ecological networks. The resulting resilience function is on a multi-dimensional manifold over the complex parameter space characterizing the system.

$$\frac{dy}{dt} = f\big(\beta(y)\big) \tag{1}$$

$$f\left(\beta(y_0)\right) = 0 \tag{2}$$

$$\frac{\partial f}{\partial y} < 0 \text{ at } y = y_0 \tag{3}$$

Equation 2 provides the resilience function as a function of the natural parameter space. The resilience function $y(\beta)$ represents possible states of the system. β is a complex parameter space that characterizes the system (system's intrinsic dynamics and structure). Separation of structure and dynamics is an essential component of this methodology.

INTERVENTION STRATEGIES TO AVOID LOSS OF RESILIENCE

Transitions take place when perturbations exceed a certain threshold. The bifurcation point at which loss of resilience occurs is unpredictable. Symbiotically connected species in ecological networks are plants and pollinators. Perturbations affect the structure of the network, Aij. Patterns of resilience depend only on the system's intrinsic dynamics (Gao, *et al.*, 2016). These investigators applied Network theory to an ecological network that considered plant–pollinator visitor interaction. Our ability to predict the system's response to external perturbations depends on our ability to separate the hidden structure from dynamics. Ecological networks can be designed either in space or in time. The dynamics of plant – the pollinator system is described by the following equation.

$$\frac{dy_i}{dt} = B_i + y_i\left(1 - \frac{y_i}{K_i}\right)\left(\frac{y_i}{C_i} - 1\right) + \sum_{j=1}^{N} A_{ij}\frac{y_i y_j}{D_i + E_i y_i + H_j y_i} \tag{4}$$

The first term represents the migration of species i from neighboring patches. In the second term, K_i, is the carrying capacity and the Allee effect is active at low population densities. In the third term, mutualistic interaction between plants and pollinators is modelled. Response function saturates for large y_i or y_j. A_{ij} was constructed using plant-pollinator relationships collected for seven ecological systems. Networks with nodes N=10 to N=1429 were studied. The system's behavior was explored under three realistic perturbations by solving equation (4) numerically.

(y^H), is a single stable fixed point. It bifurcates to a state y^H and y^L (two stable fixed points); desirable and undesirable respectively. The critical point of this bifurcation is predicted to be $\beta_{effective}^{critical} = 6.97$. The value of this parameter is determined by the system's dynamics only. It is independent of the network topology, A_{ij}. In this effective β space, all data points collapse into a single universal curve irrespective of the size and topology of the ecological network being probed or the nature of the perturbation applied. This collapse suggests the existence of universality in the resilience function irrespective of network structure and dynamics.

Complex systems are characterized by their natural parameter spaces. Hidden universal patterns of network resilience are captured by separating system's dynamics and topology (A_{ij}). Role of network topology is fully captured by β_{eff}.

$$\beta = \frac{<s^{out}s^{in}>}{s} = < s > +SH, \tag{5}$$

Where $<s> \equiv$ density of the network,

S= symmetry of the network,

H= heterogeneity of the network

A system's resilience depends on three **structural factors**

1. Density,
2. Heterogeneity, and
3. Symmetry

CONCLUSION

An advancement of meta-community theory can be achieved by designing spatially explicit models to figure out how species move between local communities. Monolayer networks would be enough for this purpose. Advancement of meta-community theory by multilayer networks where interlayer edges provide a way to develop spatially explicit models needs to be attended to. One can investigate bio-energetic flows across food webs using a temporal food web (Blonder *et al.*, 2012). Modeling relationships between a ***network of social interactions*** and a ***network of movement patterns*** to know how movement decisions are taken is highly recommended.

Network ecology provides a framework to answer questions that were difficult to answer otherwise. An integration of resilience theory with that of networks is advantageous for the advancement of ecological sciences. With this backdrop, the next chapter attempts to explore how different scenarios of climate change would affect biodiversity at all three levels; genetic, species, and ecosystems. It is a challenging task, but the benefits could be enormous. Readers have been introduced to all the necessary tools to undertake this task. Of course, methodologies are yet to be developed.

ACKNOWLEDGEMENTS

The author thanks Stephen Paul Ellner for carefully reading the first draft of the book. Implementation of his suggestions has improved flow of presentation of ideas.

REFERENCES

Bastolla, U., Fortuna, M.A., Pascual-García, A., Ferrera, A., Luque, B., Bascompte, J. (2009). The architecture of mutualistic networks minimizes competition and increases biodiversity. *Nature, 458*(7241), 1018-1020.
[http://dx.doi.org/10.1038/nature07950]

Berlow, E.L., Dunne, J.A., Martinez, N.D., Stark, P.B., Williams, R.J., Brose, U. (2009). Simple prediction of interaction strengths in complex food webs. *Proc. Natl. Acad. Sci. USA, 106*(1), 187-191.
[http://dx.doi.org/10.1073/pnas.0806823106] [PMID: 19114659]

Blonder, B., Wey, T.W., Dornhaus, A., James, R., Sih, A. (2012). Temporal dynamics and network analysis. *Methods Ecol. Evol., 3*(6), 958-972.
[http://dx.doi.org/10.1111/j.2041-210X.2012.00236.x]

Bonkoungou, E. G. Biodiversity in dry-lands: Challenges and opportunities for conservation and sustainable use, IUCN, Global Environmental Facility, Lusaka, Zambia.

Brose, U. (2010). Body-mass constraints on foraging behaviour determine population and food-web dynamics. *Funct. Ecol., 24*(1), 28-34.
[http://dx.doi.org/10.1111/j.1365-2435.2009.01618.x]

Camacho, J., Guimerà, R., Nunes Amaral, L.A. (2002). Robust patterns in food web structure. *Phys. Rev. Lett., 88*(22), 228102.
[http://dx.doi.org/10.1103/PhysRevLett.88.228102] [PMID: 12059454]

Closs, G.P., Lake, P.S. (1994). Spatial and temporal variation in the structure of an intermittent stream food – web. *Ecol. Monogr., 64*(1), 1-21.
[http://dx.doi.org/10.2307/2937053]

Contreras, S. (2021). Quiroga E. and Urzu'a, A'. "Bio - energetic traits of three keystone marine species in a food web of pristine Patagonian fjard. *J. Sea Res., 167*101984.
[http://dx.doi.org/10.1016/j.seares.2020.101984]

Cook, R.M., Hubbard, S.F. (1977). Adaptive search strategies in insect parasites. *J. Anim. Ecol., 46*(1), 115-125.
[http://dx.doi.org/10.2307/3950]

Darwin, C. (2017). *The origin of species by means of natural selection.* General Press: Daryaganj, New Delhi, Delhi, India.

Dunne, J.A., Williams, R.J., Martinez, N.D. (2002). Food-web structure and network theory: The role of connectance and size. *Proc. Natl. Acad. Sci. USA, 99*(20), 12917-12922.
[http://dx.doi.org/10.1073/pnas.192407699] [PMID: 12235364]

Fortin, M-J., Dale, M.R.T., Bascombe, C. (2021). *Network Ecology in Dynamic Landscapes Proc..* Roc. Soc. Lond, Series B..
[http://dx.doi.org/10.1098/rspb.2020.1889]

Gao, J., Barzel, B., Barabási, A.L. (2016). Universal resilience patterns in complex networks. *Nature, 530*(7590), 307-312.
[http://dx.doi.org/10.1038/nature16948] [PMID: 26887493]

Govaert, L., De Meester, L., Speak, P., Hairston, N.G. (2021). Eco - evolutionary dynamics in Freshwater Systems. *Reference Module in Earth Systems and Environmental Sciences,* (July),

[http://dx.doi.org/10.1016/B978-0-12-819166-8.00028-1]

Gross, K. (2008). Positive interactions among competitors can produce species-rich communities. *Ecol. Lett., 11*(9), 929-936.
[http://dx.doi.org/10.1111/j.1461-0248.2008.01204.x] [PMID: 18485001]

Gunderson, L.H. (2000). Ecological Resilience—In Theory and Application. *Annu. Rev. Ecol. Syst., 31*(1), 425-439.
[http://dx.doi.org/10.1146/annurev.ecolsys.31.1.425]

Holland, J.N., DeAngelis, D.L., Bronstein, J.L. (2002). Population dynamics and mutualism: functional responses of benefits and costs. *Am. Nat., 159*(3), 231-244.
[http://dx.doi.org/10.1086/338510] [PMID: 18707376]

Holling, C.S. (1996). *Engineering Resilience vs* ecological resilience: Engineering within ecological constraints.. National Academies Press.

Hubbard, S.F., Cook, R.M. (1978). Optimal foraging by parasitoid wasps. *J. Anim. Ecol., 47*(2), 593-604.
[http://dx.doi.org/10.2307/3803]

Ke'fi, S., The'bouldt, E., Eklof, A. (2017). Towards multiplex ecological networks: accounting for multiple interaction types to understand community structure and dynamics.*Adaptive Food Webs.*   Cambridge University Press.
[http://dx.doi.org/10.1017/9781316871867.008]

Kéfi, S., Berlow, E.L., Wieters, E.A., Navarrete, S.A., Petchey, O.L., Wood, S.A., Boit, A., Joppa, L.N., Lafferty, K.D., Williams, R.J., Martinez, N.D., Menge, B.A., Blanchette, C.A., Iles, A.C., Brose, U. (2012). More than a meal... integrating non-feeding interactions into food webs. *Ecol. Lett., 15*(4), 291-300.
[http://dx.doi.org/10.1111/j.1461-0248.2011.01732.x] [PMID: 22313549]

King, I.L., Li, Y. (2018). Host parasite interactions promote disease tolerance to intestinal helminth infection. *Front. Immunol., 9*, 2128.
[http://dx.doi.org/10.3389/fimmu.2018.02128] [PMID: 30298071]

Kitching, R.L. (1987). Spatial and temporal variation in food - webs in water filled tree holes. *Oikos, 48*(3), 280-288.
[http://dx.doi.org/10.2307/3565515]

Kivela, M., Arenas, A., Barthelemy, M., Gleeson, J.P., Moreno, Y., Porter, M.A. (2014). Multilayer networks. *J. Complex Netw., 2*(3), 203-271.
[http://dx.doi.org/10.1093/comnet/cnu016]

Kortsch, S., Frelat, R., Pecuchet, L., Olivier, P., Putnis, I., Bonsdorff, E., Ojaveer, H., Jurgensone, I., Strāķe, S., Rubene, G., Krūze, Ē., Nordström, M.C. (2021). Disentangling temporal food web dynamics facilitates understanding of ecosystem functioning. *J. Anim. Ecol., 90*(5), 1205-1216.
[http://dx.doi.org/10.1111/1365-2656.13447] [PMID: 33608888]

Luck, R.F., Shepard, B.M., Kenmore, P.E. (1999). *Evaluation of Biological control with experimental methods.Handbook of Biological Control.* (1st ed., pp. 225-242). San Diego, USA: Academic Press.
[http://dx.doi.org/10.1016/B978-012257305-7/50056-4]

McCann, K., Hastings, A., Huxel, G.R. (1998). Weak trophic interactions and the balance of nature. *Nature, 395*(6704), 794-798.
[http://dx.doi.org/10.1038/27427]

Mckann, K. (2007). Protecting bio - structure. *Nature, 446*(29)
[http://dx.doi.org/10.1038/446020a]

Montoya, J.M., Pimm, S.L., Solé, R.V. (2006). Ecological networks and their fragility. *Nature, 442*(7100), 259-264.
[http://dx.doi.org/10.1038/nature04927] [PMID: 16855581]

Moore, J.C., de Ruiter, P.C. (1991). Temporal and spatial heterogeneity of trophic interactions within below-

ground food webs. *Agric. Ecosyst. Environ., 34*(1-4), 371-397.
[http://dx.doi.org/10.1016/0167-8809(91)90122-E]

Naiman, R.J., Alldredge, J.R., Beauchamp, D.A., Bisson, P.A., Congleton, J., Henny, C.J., Huntly, N., Lamberson, R., Levings, C., Merrill, E.N., Pearcy, W.G., Rieman, B.E., Ruggerone, G.T., Scarnecchia, D., Smouse, P.E., Wood, C.C. (2012). Developing a broader scientific foundation for river restoration: Columbia River food webs. *Proc. Natl. Acad. Sci. USA, 109*(52), 21201-21207.
[http://dx.doi.org/10.1073/pnas.1213408109] [PMID: 23197837]

Newman, M.E.J. (2010). *Networks: an introduction..* Oxford, New York, USA: Oxford University Press.
[http://dx.doi.org/10.1093/acprof:oso/9780199206650.001.0001]

Newman, M.E.J., Girvan, M. (2004). Finding and evaluating community structure in networks. *Phys. Rev. E Stat. Nonlin. Soft Matter Phys., 69*(2), 026113.
[http://dx.doi.org/10.1103/PhysRevE.69.026113] [PMID: 14995526]

O'Gorman, E.J., Emmerson, M.C. (2009). Perturbations to trophic interactions and the stability of complex food webs. *Proc. Natl. Acad. Sci. USA, 106*(32), 13393-13398.
[http://dx.doi.org/10.1073/pnas.0903682106] [PMID: 19666606]

O'Neil, R.V., DeAngelis, D.L., Waide, J.B., Allen, T.F.H., Allen, G.E. (1986). *A Hierarchical concept of Ecosystems..* Princeton, New Jersey: Princeton University Press.

Ollivier, M., Lesieur, V., Raghu, S., Martin, J.F. (2001). "Characterizing ecological networks to support risk assessment in classical biological control of weeds", *Curr. Opin. Insec. Sci.*, vol. 38, no., pp. 40 - 47, April 2020. *Nature*

Pilosof, S., Porter, M.A., Pascual, M., Kéfi, S. (2017). The multilayer nature of ecological networks. *Nat. Ecol. Evol., 1*(4), 0101.
[http://dx.doi.org/10.1038/s41559-017-0101]

Poisot, T., Canard, E., Mouillot, D., Mouquet, N., Gravel, D. (2012). The dissimilarity of species interaction networks. *Ecol. Lett., 15*(12), 1353-1361.
[http://dx.doi.org/10.1111/ele.12002] [PMID: 22994257]

Rudolf, V.H.W., Lafferty, K.D. (2011). Stage structure alters how complexity affects stability of ecological networks. *Ecol. Lett., 14*(1), 75-79.
[http://dx.doi.org/10.1111/j.1461-0248.2010.01558.x] [PMID: 21114747]

Saavedra, S. (2009). F. Reed – Tsochas and B. Uzzi, "A simple model of bipartite cooperation for ecological and organization networks". *Nature, 457*(7228), 463-466.
[http://dx.doi.org/10.1038/nature07532] [PMID: 19052545]

Sait, S.M., Begon, M., Thompson, D.J., Harvey, J.A., Hails, R.S. (1997). Factors affecting host selection in an insect host–parasitoid interaction. *Ecol. Entomol., 22*(2), 225-230.
[http://dx.doi.org/10.1046/j.1365-2311.1997.t01-1-00051.x]

Scheffer, M., Carpenter, S., Foley, J.A., Folke, C., Walker, B. (2001). Catastrophic shifts in ecosystems. *Nature, 413*(6856), 591-596.
[http://dx.doi.org/10.1038/35098000] [PMID: 11595939]

Schoenly, K., Cohen, J.E. (1991). Temporal variation in food - web structure: 16 empirical cases. *Ecol. Monogr., 61*(3), 267-298.
[http://dx.doi.org/10.2307/2937109]

Starr, D.J., Cline, T.W. (2002). A host–parasite interaction rescues Drosophila oogenesis defects. *Nature, 418*(6893), 76-79.
[http://dx.doi.org/10.1038/nature00843] [PMID: 12097909]

Trojelsgaard, K., Jordano, P., Karstensen, D. W., Olsen, J. M. (2015). Geographical variations in mutualistic networks: similarity, turn over and partner fidelity *Proc. Roy. Soc. B.*
[http://dx.doi.org/10.1098/rspb.2014.2925]

Ulrich, W., Soliveres, S., Maestre, F.T., Gotelli, N.J., Quero, J.L., Delgado-Baquerizo, M., Bowker, M.A., Eldridge, D.J., Ochoa, V., Gozalo, B., Valencia, E., Berdugo, M., Escolar, C., García-Gómez, M., Escudero, A., Prina, A., Alfonso, G., Arredondo, T., Bran, D., Cabrera, O., Cea, A.P., Chaieb, M., Contreras, J., Derak, M., Espinosa, C.I., Florentino, A., Gaitán, J., Muro, V.G., Ghiloufi, W., Gómez-González, S., Gutiérrez, J.R., Hernández, R.M., Huber-Sannwald, E., Jankju, M., Mau, R.L., Hughes, F.M., Miriti, M., Monerris, J., Muchane, M., Naseri, K., Pucheta, E., Ramírez-Collantes, D.A., Raveh, E., Romão, R.L., Torres-Díaz, C., Val, J., Veiga, J.P., Wang, D., Yuan, X., Zaady, E. (2014). Climate and soil attributes determine plant species turnover in global drylands. *J. Biogeogr., 41*(12), 2307-2319.
[http://dx.doi.org/10.1111/jbi.12377] [PMID: 25914437]

Vandermeer, J., Evans, M.A., Foster, P., Höök, T., Reiskind, M., Wund, M. (2002). Increased competition may promote species coexistence. *Proc. Natl. Acad. Sci. USA, 99*(13), 8731-8736.
[http://dx.doi.org/10.1073/pnas.142073599] [PMID: 12070354]

Walker, B.H., Ludwig, D., Holling, C.S., Peterman, R.M. (1981). Stability of semi - arid savanna grazing systems. *J. Ecol., 69*(2), 473-498.
[http://dx.doi.org/10.2307/2259679]

Warren, L., Manos, P.D., Ahfeldt, T., Loh, Y.H., Li, H., Lau, F., Ebina, W., Mandal, P.K., Smith, Z.D., Meissner, A., Daley, G.Q., Brack, A.S., Collins, J.J., Cowan, C., Schlaeger, T.M., Rossi, D.J. (2010). Highly efficient reprogramming to pluripotency and directed differentiation of human cells with synthetic modified mRNA. *Cell Stem Cell, 7*(5), 618-630.
[http://dx.doi.org/10.1016/j.stem.2010.08.012] [PMID: 20888316]

Wilson, E.O. (2014). *The meaning of human existence..* London: W. W. Norton and Company.

Winemiller, K.O., Jepsen, D.B. (1998). Effects of seasonality and fish movement on tropical river food webs. *J. Fish Biol., 53*(sA), 267-296.
[http://dx.doi.org/10.1111/j.1095-8649.1998.tb01032.x]

Winemiller, K.O., Polis, G.A. (1996). *Food - Webs: what do they tell us.Food webs: integration of patterns and dynamics.* (pp. 298-312). New York, USA: Springer.
[http://dx.doi.org/10.1007/978-1-4615-7007-3_29]

Ylönen, H., Haapakoski, M., Sievert, T., Sundell, J. (2019). Voles and weasels in the boreal Fennoscandian small mammal community: what happens if the least weasel disappears due to climate change? *Integr. Zool., 14*(4), 327-340.
[http://dx.doi.org/10.1111/1749-4877.12388] [PMID: 30811858]

CHAPTER 2

Biodiversity and Climate Change: The Missing Link

Abstract: Changes in an organism's DNA can influence all aspects of its life. Mutations serve as raw material for genetic variability and its evolution. These are caused by high-energy radiation. Chemical substances present in the environment are other potential causative agents. They may also occur during DNA replication. Radiation damage has increased many folds after the advent of cellular telephony. Genes are DNA molecules. These molecules are distributed on the chromosomes of individuals or populations of a species. Some populations grow faster than others. Why? The chapter tries to find an answer to it.

Population increase has been observed in some countries and for others, a decrease has been detected. It has been predicted that the human population will increase to 8.5 billion in 2030 from 7.8 billion in 2020. The exploitation of natural resources would increase accordingly. **World Economic Forum** reports that research conducted at the Swiss Re Institute has pointed out that an 18 percent reduction in GDP is achievable by 2050 if the **Global temperature rise** is restricted to 3.2^0 C. About 16 to 29% reduction in CO_2 emissions would lead to a less dangerous climate change provided population growth is slowed down. If factors contributing to Global warming are managed in such a way that the Global temperature rises by 1.5^0 C, **Sustainable Development Goals** are achievable. Net–zero emissions targeted to be achieved by 2050 are not feasible as international agreements are not honored.

Keywords: Biodiversity, Diversity, Extinction, Genetic variability, Meiosis, Niche differentiation, Niche dimensionality, Nonlinear oscillations, Passenger pigeon, Plant-pollinator systems, Representative concentration pathways.

INTRODUCTION

Populations of a species are distributed in space (geographical locations). Species migrate from one patch to another in a fragmented landscape. The genomic buildup of individuals varies both in space and time. This chapter discusses factors that cause loss of biodiversity at all three levels: *Genetic, Species, and Ecosystems.*

GENETIC DIVERSITY

It has three components

1. Genetic make - up of a species,
2. Genetic variability,
3. Genetic diversity

The genetic makeup of a species is determined by how many genes are there on the chromosomes of an individual plant or animal species. The same applies to viruses that associate themselves with these species. Genetic diversity is the total number of genetic characteristics in the genetic makeup of a species. It is different from genetic variability in the sense that genetic variability describes the tendency of genetic characteristics to vary. Genetic diversity is a measure of the potential of a population to survive in fluctuating environments. The genetic variation can be caused by mutation, random mating, random fertilization, and recombination between homologous chromosomes during *meiosis*. Meiosis reshuffles alleles within an organism's offspring. Random fertilization occurs in plant-pollinator systems. The Kimura - Crow theory (Kimura & Crow, 1970) suggests that evolution at the molecular level occurs by random drift of neutral mutations.

Passenger pigeons are blue, long-tailed, fast, and graceful. These migratory birds are endemic to North America. Flocks a mile wide flying overhead from 7. 3 a.m. in the morning and 4 p.m. in the afternoon are a few characteristics. Pigeons migrate in the early spring from the south to their breeding areas in New England, New York, Ohio, and the southern Great Lakes. The death of a passenger pigeon, Martha, at Cincinnati Zoo at the age of 29 prompted zoologists to take measures to conserve these species of pigeons, which were hunted to extinction. Species with low genetic variation are more prone to extinction long after their population size is recovered. Murray *et al.* (2017) discussed the rise and fall of genetic diversity in passenger pigeons. The authors studied the genomes of four passenger pigeon samples from different localities within its range. The interplay between passenger pigeon population size, genomic structure and recombination, and natural selection was explored. The genetic diversity provided limited avenues for the bird to respond to human pressures, which dumped it into oblivion (Hung *et al.*, 2014).

Passenger pigeons could not switch over to living in smaller flocks. Lee *et al.* (2010) examined the rate of loss of genetic diversity of the bird to figure out why it could not survive in a few small populations. Investigators explored the demography and genetic structure of winter flocks in a small passerine, the vinous throated parrotbill, *paradoxornis webbianus*. The objective was to determine the match between observed demography and the genetic structure of winter flocks and the consequences of kin structure for the risk of inbreeding during the breeding season. With 600 to 120 individuals, there was no deviation from parity in the sex ratio. The survival of adults was moderately low (Berner & Grubb, 1985; Lee *et al.*, 2010; Pulliman, 1973).

SPECIES DIVERSITY

There exist ~ 8.7 million total species on this earth. There are a total of 6.5 million species on land (forests, mangroves, marshes, wetlands, *etc.*) and 2.2 million in oceans. One square kilometer of a forest can have a hundred different tree species; broadleaf trees, mosses, ferns, and orchids all thrive in rainforests. Densely growing trees and their branches block the sun from penetrating the understory. A variety of birds, bats, and other animals live on them.

It is the number of species present in a given community. It represents 1) a number of species in a community, and 2) the evenness of species abundances; *i.e.*, its richness and evenness. Species evenness is the variation in the abundance of individuals per species in a community. The relationship between species richness and evenness in plant communities was investigated by Zhang, *et al.* (Zhang, *et al.*, 2012). A total number of 30 sampling quadrants of size 0.5m×0.5m were laid out along two transects at each meadow. The authors found a negative correlation between richness and evenness in these communities along the succession gradient at the sampling site. The authors explored the relationship by varying the scale in the range (0.5m×0.5m-10m×10m). They found that niche differentiation and spatial scale effects cooperate to maintain high species richness in sub-alpine meadows communities. Niche differentiation is the process by which competing species use the environment for coexistence.

Variation within species is measured using differences in the base sequence of DNA or amino acid sequence of proteins. This method is known as sequence-based identification of biodiversity. Biodiversity within a community can be measured using species richness and an index of diversity (Creer, *et al.*, 2016).

ECOSYSTEM DIVERSITY

It has two aspects: 1) habitat, and 2) community. Ecosystems are both ***natural*** and ***artificial***. Natural ecosystems are of four types: 1) Forest, 2) Grassland, 3) Aquatic, and 4) Desert. Artificial ecosystems are of three types: terrestrial, microbial, and aquatic.

STABILITY AND DIVERSITY OF ECOSYSTEMS

Ives and Carpenter (2007) have argued that the diversity–stability debate can only be understood in the context of how environmental drivers affect both interacting species and their environment. The strength of interaction among species, the topology of food-webs, and sensitivies of species to different kinds of environmental perturbations are a few key factors. Field studies have focused on diversity because it is easy to measure and provides a method to assess the loss of biodiversity. The relationship between stability and diversity has been a subject of intense debate primarily because contradictions were widespread in theoretical predictions. Moreover, empirical studies were inconsistent. Theoreticians and empiricists found it difficult to resolve disagreements. There is an urgent need to identify hotspots of biodiversity and debate and discuss the subject to evolve a methodology to provide a basis for efficient methods and strategies to conserve it at all three levels: *genetic, species, and ecosystems.*

NICHE DIMENSIONALITY AND GRASSLAND DIVERSITY

Harpole and Tilman (2007) reported outcomes of certain field experiments to test theories of high diversity. The prediction was that the high diversity of coexisting competitors could result from greater niche dimensionality; *i.e.*, a large number of limited resources, as a large number of species can coexist when limited by just one or two factors *in situ*ations when species have just one or two factors with appropriate trade-offs. In these experiments, investigators manipulated the number of added limiting soil resources (soil moisture, nitrogen, phosphorous, and base cations). An increased supply of multiple limiting resources can reduce niche dimensionality.

Niche dimensionality plays a key role in the theoretical explanation of observed biodiversity. Niches are characterized by *limiting factors*. Harpole *et al.* (2016) found that elevated resource supply reduced niche dimensionality and diversity and increased both productivity and compositional turnover. The authors stress the importance of understanding dimensionality in ecological systems that have suffered loss of diversity in response to multiple global agents of change.

HOT SPOTS OF BIODIVERSITY

It is a bio-geographic region with a significant number of endangered species. Endangerment is caused by several factors; *e.g.*, the development of the real estate, construction of dams and bridges, hosting of cultural festivals, *etc.* According to the Myers 2000 edition of the hotspot map, a region must meet two strict criteria: 1) it must contain at least 0.5% or 1500 species of vascular plants as endemics, and 70% of its primary vegetation is already extinct. Total 36 areas around the world satisfy these criteria. The identification of hotspots is based on quantitative criteria with a few subjective considerations. The information of a number of taxa present in an ecosystem is just not sufficient for devising conservation strategies. The science policy partnerships in an interdisciplinary framework have the potential to set conservation priorities. Three types of hotspots of biodiversity have been delineated: 1) hotspots of endemic species that are found only in that area, 2) hotspots of endangered species, and 3) hotspots of richness.

THE CONNECTION

Genetic diversity within species increases with environmental variability. Different genes are at an advantage at different times and places in variable environments. Therefore, at any given time, there are several genes. On contrary to this, in static environments, a small number of genes would spread at the cost of others, thus lowering the genetic diversity. Genetic diversity can increase in communities depending on the fact that how closely related the species are. The number of species present is also important, but the relatedness of species within communities is more important. Species diversity receives more attention as the other two diversities demand huge efforts, complex measurements, and funding. Moreover, species are distributed units of diversity. Each species has been assigned a particular role in the ecosystem. The number of different species in an ecosystem is a measure of the health of the ecosystem. The term ecosystem refers to all levels greater than species, associations, and communities. This is the least understood level. A difficulty often faced in investigating communities is the absence of a sharp boundary; *e.g.*, a lake having a sharp boundary with a deciduous forest will shift gradually to grasslands or to a coniferous forest. The lack of sharp boundaries gives rise to the existence of "open communities".

Properties of communities are special to that level. Characteristics of communities are unique and cannot be extrapolated from the species level. These characteristics could be the food chain and the species at those levels, guilds (functionally similar species in a community), and interactions of other types. Genomic and ecological analyses indicate that the passenger pigeons experienced dramatic population fluctuations which increased their vulnerability to human exploitation. High-throughput DNA analysis combined with ecological niche modeling can provide evidence that would enable us to identify and assess factors that led to the surprisingly rapid demise of the passenger pigeon. Analysis of DNA recovered from museum specimens reveals low genetic diversity of the passenger pigeon population. Its genome shows the effects of natural selection in a large population.

Adaptive social behavior has demographic consequences. How do birds choose mates? *De novo* evolution of cooperation in plants, birds, and animal societies (Szekely, *et al.*, 2010) has been discovered. The advantages of participating in mixed-species foraging groups have been analyzed in the light of two models that investigators developed for the purpose. Changes in nesting microhabitats after predation are well known. There is not much work that addresses proactive avoidance of areas with many nest predators even though this kind of avoidance is widespread. Lima (2009) has discussed behavioral and reproductive flexibility under the risk of predation in a breeding bird. Individual birds adapt anti-predator strategies based on the nest predators actually present in the area. It has been established that breeding birds have the capacity to assess and respond, over an ecological time scale, to changes in the risk of predation to both themselves and their eggs or nests. Ulrika and Wong (2012) discussed how species respond to changes in their habitat and the climate.

CONSERVATION OF BIODIVERSITY

A large number of biologists are of the view that we are in the midst of a mass extinction. It means that 75% of the species would be lost over a short geological time scale (Raup, 1994). It is important to conserve plant species as most of the medicines are derived from plants. Opiate pain relievers are derived from poppies, aspirin from willows, and quinine from the Chinchona tree. Trees provide the raw material for making homes, furniture, and paper products. The rosy periwinkle (Vince rosea) and pacific yew (Taxa brevififolia), which provide substances used in the treatment of cancer through chemotherapy (Kearns, 2010) need to be protected.

Extinction and Evolution

Extinction is likely when killing stress is outside natural selection. Mass extinctions trigger restructuring of the biosphere when some successful groups are eliminated. It can not be predicted which species will be victims of an extinction event.

Mutation and Evolution

Changes in an organism's DNA can influence all aspects of its life. Mutations serve as raw material for genetic variability and evolution. These are caused by high-energy radiation, and the chemical substances present in the environment. They may also occur during DNA replication. The human genome contains over 3.1 billion bases of DNA. Cell division demands faithful replication. Chromosomal aberrations are larger-scale mutations that can occur during **meiosis** in unequal crossing–over events, slippage; and slippage during DNA recombination. Genes or whole chromosomes can be substituted or deleted. Mutations cause slow changes in allele frequency.

Nonlinear Oscillations

Vandermeer (2006) provided a general framework for conserving biodiversity based on his analysis of classic equations of ecology. The author explored *phase coordination* in two coupled consumer-resource systems with niche overlap in consumers. An understanding of the nature of phase coordination sets a platform for an understanding of *exclusion* and *coexistence*. A species invading the environment of two consumer-resource oscillators gets coupled to the two original oscillators.

Synchronization (Strogatz, 2000) of chaotic oscillations (Rai, 2013; Van dermeer, *et al.*, 2006) facilitate invasion. This opportunity of invasion is unpredictable as communities are embedded in stochastic environments. Vandermeer and a collaborator explored *competitive co-existence* through *intermediate polyphagy* (Vandermeer & Pascual, 2006). Readers are advised to refer to this literature for a lucid description of competitive coexistence.

MANAGEMENT OF ECOLOGICALLY SENSITIVE HOTSPOTS OF BIODIVERSITY

An ecologically sensitive hotspot of biodiversity is one that has several endemic species, belonging to different taxa, that are endangered. Western Ghats were declared as an ecologically sensitive hotspot in 1988. This region has a number of protected areas including two biosphere reserves, 14 national parks, and several wildlife sanctuaries. A few of them have been declared as reserve forests. Other hotspots are the Himalayas and the Indo-Burma region. The International Union for Conservation of Nature works on biodiversity, climate change, energy, and human livelihoods (Allen, *et al.*, 2011). Identification of key biodiversity areas (KBA), which make significant contributions to the global persistence of diversity, is based on the following criteria. Species, facing a high risk of extinction and having a major share in the global persistence of biodiversity at the genetic and species levels, characterize KBAs.

KBAs are defined based on the following four criteria:

1. Vulnerable species,
2. Endangered species, and
3. Critically endangered species designated by the IUCN red list.

One of the criteria is based on the presence of species in range-limited and biome–restricted areas. The other criterion aims at identifying the roosting of breeding sites containing a congregation of individuals of species as part of their life histories (Eken *et al.*, 2004). Global thresholds are yet to be set for many taxa. Globally significant populations can be identified based on the information contained in the previous studies (Night, *et al.*, 2007).

Richness and Endemism are two key factors for setting conservation measures for biodiversity. The area-based approach can be applied to any geographical scale and is considered to be one of the best approaches for the maintenance of biological diversity. Ceballos and Ehrlich (2009) examined the overlap between the top 2.5% hotspots of each type. Total 16% of species were found in all three types. They also examined the overlap within the hotspots. They looked for specific areas within the hotspots which contained a higher number of unique species. Researchers proposed "optimism algorithms" to help guide conservation efforts. These techniques help figure out the maximum number of species that can be protected by the minimum number of sites. Since current extinction rates are far higher than the rates during previous mass extinctions, it is estimated that Earth's biota is entering into sixth mass extinction.

Biodiversity conservation is executed in two forms: *in situ* and *ex-situ*. Protected areas belong to the first category. National parks, bird sanctuaries, and Biosphere reserves are a few examples of protected areas. *Ex–situ* conservation involves seed banks, field gene banks, and cryopreservation. Botanical gardens, zoological parks, aquaria, and arboreta are examples of Biosphere reserves.

CONSERVATION OF BIODIVERSITY AND CLIMATE CHANGE

Mooney *et al.* (2009) discussed the relationship between biodiversity and climate change. Schwalm, *et al.* (2017) captured trends in the historical records of the recovery of ecosystems from drought using three state-of-the-art sets of gross primary productivity; the amount of atmospheric carbon dioxide that is fixed in ecosystems by photosynthesis. These investigators analyzed the data and found that the total area of ecosystems had increased during the 20th century. The authors were able to identify areas where the recovery from drought was delayed. The intergovernmental panel on climate change focuses on mean climate changes rarely providing the frequency of the extreme events. The statistics of time lapses in recovery are not available. There has been increasing attention on the time scale of the recovery processes as part of impact assessments. Drought impacts both human and natural systems. It negatively impacts the land carbon sink. More frequent droughts are expected in the 21st century. The time taken by an ecosystem to return to its pre-drought functional state, is a critical measure of drought impact (Semviratne & Ciais, 2017). Analyzing three independent data sets of drought recovery and its spatio-temporal patterns at the global level, investigators conclude that

1. Recovery is longest in the tropics and high latitudes,
2. Drought impacts have improved over the twentieth century.

CLIMATE CHANGE

Climate Change can be tagged to a less dangerous level if population growth is slowed down (O' Neill *et al.*, 2010). Is it possible to slow down the population growth of the world population? What should be done by countries to achieve this goal? Climate modelers have developed Representative concentration pathways to figure out trajectories of Climate Change.

REPRESENTATIVE CONCENTRATION PATHWAYS (RCPS)

RCPs are a set of four new pathways developed for the climate modeling community. These are valid up to 2100. Radiative forcing values for 2010 have been predicted to be 2.6 to 8.5 Wm^2. It is the outcome of a collaborative exercise between integrated assessment modelers, climate modelers, and emission inventory experts. Socio-economic and emission scenarios are developed based on the following variables:

1. Socioeconomic change,
2. Technological Change,
3. Land Use Change,
4. Energy,
5. Emission of Green House Gases, and
6. Air pollutants.

0.5×0.5 degree spatial resolution per sector was used for greenhouse gases. Coarser resolution is used for well-mixed gases. The following plays a vital role in the climatic drivers' impact on ecosystem dynamics.

1. Land Use (Wise *et al.*, 2009),
2. Emission of air pollutants (Greenhouse gases).

Extended Concentration Pathways (ECPs) that extend up to 2300 are a new set of scenarios. The Intergovernmental Panel on Climate Change (IPCC) prompted the scientific communities to develop a new set of scenarios to facilitate future assessments (IPCC, 2007). RCPs, selected from the existing literature on the basis of emissions and associated concentration levels, do not provide information on socio-economic parameters (Van Vuuren, *et al.*, 2011).

The population and GDP pathways are shown in Fig. (**1**). This shows UN population projections and the 90th percentile range of GDP scenarios in the literature on greenhouse gas emissions. In contrast to other scenarios, RCP8.5 was based on a revised version of the SREAS A2 scenario, which is characterized by high population growth and lower income in developing countries. RCPs are appropriate conditions in interpreting projections for scenario elements that are indirectly coupled to radiative forcing targets; *e.g.*, 1) land use/ land cover, 2) socio-economic parameters, 3) and emissions of short-lived species to some extent.

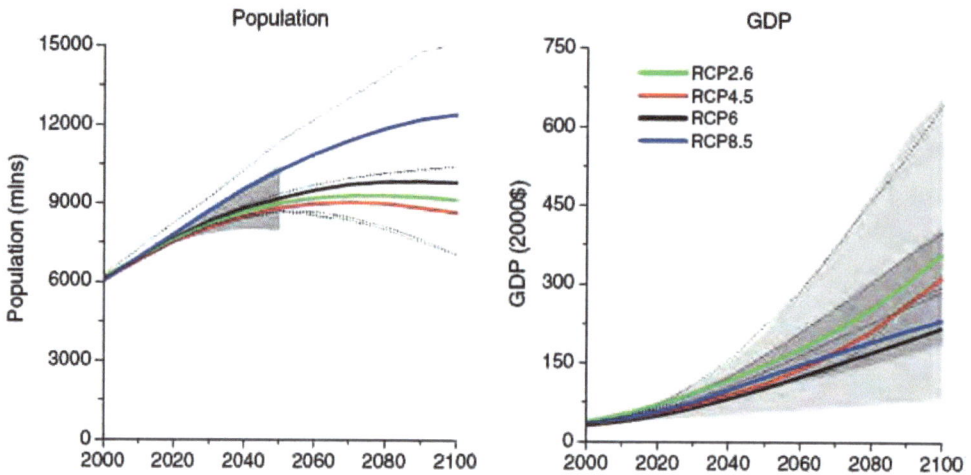

Fig. (1). Projections of population and GDP pathways (Source: Van Vuuren, *et al.*, *Climatic Change*, vol. 109, pp. 5 – 31, 2011).

Radiative forcing pathways consistent with those in the current literature are derived on the basis of internally consistent socio-economic assumptions. RCPs, as a set, do not provide internal logic. They do not cover the entire range of socio-economic trajectories that exist in the literature. A community effort is needed to define the ***socio-economic dimension*** that would complement RCPs.

Hampe and Petit (2005) studied long-term stores of genetic diversity of species from low latitudes and foci of speciation by analyzing recent findings from fossil records, –phytogeography, and ecology. The analysis suggested that the rear edge species were vital more than expected survival and evolution of biota. Commonly recommended conservation practices might be of little use or counterproductive for **rear edge populations**.

PREDICTIONS

Following are the predictions:

1. Pole ward range shifts of numerous taxa, communities, and ecosystems worldwide.
2. The environmental responses to this shift would be determined by population responses.

CONCLUSION

Ecosystems have been under stress to deliver essential ecosystem services for a quite long time. Climate change adds to this stress in such a way that most of the rivers have been restructured, oceans have been severely altered, and depleted coral reefs are about to disappear as functional ecosystems. More than half of the land surface is devoted to livestock and crop agriculture ignoring the loss of ecosystem services as a consequence. Climate change would cause disruption in the ecosystem base. This implies extinctions would take place within vulnerable habitats. Conditions where migrations are necessary for survival, but no pathways are available for successful movement in this fragmented world, would be a common feature.

Homo sapiens are a **keystone species**; species which are affected most by other species in the ecosystem. Of course, these are the ones that influence other species profoundly. The next chapter puts these species at the center of all ecological considerations. This is how yet another sub-discipline of ecological sciences emerges. It is named as **human ecology**. Readers are introduced to various tenets of human ecology. An outline of how ecological sciences can be integrated with neurosciences is provided. Such integration would broaden the scope of ecological sciences.

ACKNOWLEDGEMENTS

The author thanks Ranjit Kumar Upadhyay for helpful comments and Stephen Paul Ellner who read the first draft of the textbook carefully.

REFERENCES

Allen, D.J., Molur, S., Daniel, B.A. (2011). *The status and distribution of freshwater biodiversity in the eartern Himalaya, IUCN report.*. IUCN.

Ceballos, G., Ehrlich, P.R. (2002). Mammal population losses and the extinction crisis. *Science, 296*(5569), 904-907.
[http://dx.doi.org/10.1126/science.1069349] [PMID: 11988573]

Ceballos, G., Ehrlich, P.R., Barnosky, A.D., García, A., Pringle, R.M., Palmer, T.M. (2015). Accelerated modern human–induced species losses: Entering the sixth mass extinction. *Sci. Adv., 1*(5), e1400253. [http://dx.doi.org/10.1126/sciadv.1400253] [PMID: 26601195]

Creer, S., Deiner, K., Frey, S., Porazinska, D., Taberlet, P., Thomas, W.K., Potter, C., Bik, H.M. (2016). The ecologist's field guide to sequence□based identification of biodiversity. *Methods Ecol. Evol., 7*(9), 1008-1018. [http://dx.doi.org/10.1111/2041-210X.12574]. [http://dx.doi.org/10.1111/2041-210X.12574]

Eken, G., Bennun, L., Brooks, T.M., Darwall, W., Fishpool, L.D.C., Foster, M., Knox, D., Langhammer, P., Matiku, P., Radford, E., Salaman, P., Sechrest, W., Smith, M.L., Spector, S., Tordoff, A. (2004). Key Biodiversity Areas as the site conservation targets. *Bioscience, 54*(12), 1110-1118. [http://dx.doi.org/10.1641/0006-3568(2004)054[1110:KBAASC]2.0.CO;2]

Hampe, A., Petit, R.J. (2005). Conserving biodiversity under climate change: the rear edge matters. *Ecol. Lett., 8*(5), 461-467. [http://dx.doi.org/10.1111/j.1461-0248.2005.00739.x] [PMID: 21352449]

Harpole, W.S., Sullivan, L.L., Lind, E.M., Firn, J., Adler, P.B., Borer, E.T., Chase, J., Fay, P.A., Hautier, Y., Hillebrand, H., MacDougall, A.S., Seabloom, E.W., Williams, R., Bakker, J.D., Cadotte, M.W., Chaneton, E.J., Chu, C., Cleland, E.E., D'Antonio, C., Davies, K.F., Gruner, D.S., Hagenah, N., Kirkman, K., Knops, J.M.H., La Pierre, K.J., McCulley, R.L., Moore, J.L., Morgan, J.W., Prober, S.M., Risch, A.C., Schuetz, M., Stevens, C.J., Wragg, P.D. (2016). Addition of multiple limiting resources reduces grassland diversity. *Nature, 537*(7618), 93-96. [http://dx.doi.org/10.1038/nature19324] [PMID: 27556951]

Harpole, W.S., Tilman, D. (2007). Grassland species loss resulting from reduced niche dimension. *Nature, 446*(7137), 791-793. [http://dx.doi.org/10.1038/nature05684] [PMID: 17384633]

Hung, C.M., Shaner, P.J.L., Zink, R.M., Liu, W.C., Chu, T.C., Huang, W.S., Li, S.H. (2014). Drastic population fluctuations explain the rapid extinction of the passenger pigeon. *Proc. Natl. Acad. Sci. USA, 111*(29), 10636-10641. [http://dx.doi.org/10.1073/pnas.1401526111] [PMID: 24979776]

Ives, A.R., Carpenter, S.R. (2007). Stability and diversity of ecosystems. *Science, 317*(5834), 58-62. [http://dx.doi.org/10.1126/science.1133258] [PMID: 17615333]

J. H. vandermeer, H. Liere and B. Lin, "Predation pressure on species packing on a resource gradient: insights from Nonlinear Dynamics". *J. Theor. Popl. Biology, 69*, 395-408. [http://dx.doi.org/10.1016/j.tpb.2005.11.008]

Kearns, C. (2010). Conservation of Biodiversity. *Nature Education Knowledge, 3*(10), 7. [PMID: 21092321]

Khinchi, S. (2016). Shyam and M. Tanwar, Global Climatic Change and Biodiversity. V. L. Media Solutions. Berner, T.O., Grubb, T.C., Jr (1985). Grubb, "An experimental analysis of mixed – species flocking in birds of deciduous woodlands". *Ecology, 66*(4), 1229-1236. [http://dx.doi.org/10.2307/1939176].

Kimura, M., Crow, J.F. (1970). *An Introduction to Population Genetics Theory.*. New York: Harper and Row.

Knight, A.T., Smith, R.J., Cowling, R.M., Desmet, P.G., Faith, D.P., Ferrier, S., Gelderblom, C.M., Grantham, H., Lombard, A.T., Maze, K., Nel, J.L., Parrish, J.D., Pence, G.Q.K., Possingham, H.P., Reyers, B., Rouget, M., Roux, D., Wilson, K.A. (2007). Improving the key biodiversity areas approach for effective conservation planning. *Bioscience, 57*(3), 256-261. [http://dx.doi.org/10.1641/B570309]. [http://dx.doi.org/10.1641/B570309]

Lee, J-W., Simeoni, M., Burke, T., Hatchwell, B.J. (2010). The consequences of winter flock demography for genetic structure and inbreeding risk in vinous-throated parrotbills, Paradoxornis webbianus. *Heredity, 104*(5), 472-481. [http://dx.doi.org/10.1038/hdy.2009.135] [PMID: 19812618]

Lima, S.L. (2009). Predators and the breeding bird: behavioral and reproductive flexibility under the risk of

predation. *Biol. Rev. Camb. Philos. Soc., 84*(3), 485-513.
[http://dx.doi.org/10.1111/j.1469-185X.2009.00085.x] [PMID: 19659887]

Marchese, C. (2015). Biodiversity hotspots: A shortcut for a more complicated concept. *Glob. Ecol. Conserv., 3*, 297-309. [http://dx.doi.org/10.1016/j.gecco.2014.12.008].
[http://dx.doi.org/10.1016/j.gecco.2014.12.008]

Mooney, H., Larigauderie, A., Cesario, M., Elmquist, T., Hoegh-Guldberg, O., Lavorel, S., Mace, G.M., Palmer, M., Scholes, R., Yahara, T. (2009). Biodiversity, climate change, and ecosystem services. *Curr. Opin. Environ. Sustain., 1*(1), 46-54. [http://dx.doi.org/10.1016/j.cosust.2009.07.006].
[http://dx.doi.org/10.1016/j.cosust.2009.07.006]

Murray, G.G.R., Soares, A.E.R., Novak, B.J., Schaefer, N.K., Cahill, J.A., Baker, A.J., Demboski, J.R., Doll, A., Da Fonseca, R.R., Fulton, T.L., Gilbert, M.T.P., Heintzman, P.D., Letts, B., McIntosh, G., O'Connell, B.L., Peck, M., Pipes, M.L., Rice, E.S., Santos, K.M., Sohrweide, A.G., Vohr, S.H., Corbett-Detig, R.B., Green, R.E., Shapiro, B. (2017). Natural selection shaped the rise and fall of passenger pigeon genomic diversity. *Science, 358*(6365), 951-954.
[http://dx.doi.org/10.1126/science.aao0960] [PMID: 29146814]

Myers, N., Mittermeier, R.A., Mittermeier, C.G., da Fonseca, G.A.B., Kent, J. (2000). Biodiversity hotspots for conservation priorities. *Nature, 403*(6772), 853-858.
[http://dx.doi.org/10.1038/35002501] [PMID: 10706275]

O'Neill, B.C., Dalton, M., Fuchs, R., Jiang, L., Pachauri, S., Zigova, K. (2010). Global demographic trends and future carbon emissions. *Proc. Natl. Acad. Sci. USA, 107*(41), 17521-17526.
[http://dx.doi.org/10.1073/pnas.1004581107] [PMID: 20937861]

Pulliam, H.R. (1973). On the advantages of flocking. *J. Theor. Biol., 38*(2), 419-422.
[http://dx.doi.org/10.1016/0022-5193(73)90184-7] [PMID: 4734745]

Rai, V. (2013). *Spatial Ecology: Patterns and Processes.*. Sharjah, UAE: Bentham Science Publishers.

Raup, D.M. (1994). The role of extinction in evolution. *Proc. Natl. Acad. Sci. USA, 91*(15), 6758-6763.
[http://dx.doi.org/10.1073/pnas.91.15.6758] [PMID: 8041694]

Schwalm, C.R., Anderegg, W.R.L., Michalak, A.M., Fisher, J.B., Biondi, F., Koch, G., Litvak, M., Ogle, K., Shaw, J.D., Wolf, A., Huntzinger, D.N., Schaefer, K., Cook, R., Wei, Y., Fang, Y., Hayes, D., Huang, M., Jain, A., Tian, H. (2017). Global patterns of drought recovery. *Nature, 548*(7666), 202-205.
[http://dx.doi.org/10.1038/nature23021] [PMID: 28796213]

Seneviratne, S.I., Ciais, P. (2017). Trends in ecosystem recovery from drought. *Nature, 548*(7666), 164-165.
[http://dx.doi.org/10.1038/548164a] [PMID: 28796216]

Strogatz, S. (2000). *Sync: The emerging science of spontaneous order.*. New York: Hyperion Press.

Szekely, T., Moore, A.J., Komdeur, J. (2010). *Social Behavior: genes, ecology and evolution.* (p. 30). Cambridge, U. K.: Cambridge University Press. [http://dx.doi.org/10.1017/CBO9780511781360]
[http://dx.doi.org/10.1017/CBO9780511781360]

Ulrika, C., Bob, B.M. Wong (editors) *Behavioral responses to a changing world*, Oxford University Press: Oxford, U. K., 2012.

van Vuuren, D.P., Edmonds, J., Kainuma, M., Riahi, K., Thomson, A., Hibbard, K., Hurtt, G.C., Kram, T., Krey, V., Lamarque, J.F., Masui, T., Meinshausen, M., Nakicenovic, N., Smith, S.J., Rose, S.K. (2011). The representative concentration pathways: an overview. *Clim. Change, 109*(1-2), 5-31.
[http://dx.doi.org/10.1007/s10584-011-0148-z].
[http://dx.doi.org/10.1007/s10584-011-0148-z]

Vandermeer, J., Pascual, M. (2006). Competitive coexistence through intermediate polyphagy. *Ecol. Complex., 3*(1), 37-43. [http://dx.doi.org/10.1016/j.ecocom.2005.05.005].
[http://dx.doi.org/10.1016/j.ecocom.2005.05.005]

Vandermeer, J., Stone, L., Blasius, B. (2001). Categories of chaos and fractal basin boundaries in forced

predator–prey models. *Chaos Solitons Fractals,*   *12*(2), 265-276. [http://dx.doi.org/10.1016/S0960-0779(00)00111-9].
[http://dx.doi.org/10.1016/S0960-0779(00)00111-9]

Vandermeer, J.H. (2006). Oscillating populations and maintenance of biodiversity. *Bioscience, 56*(12), 967-975.
[http://dx.doi.org/10.1641/0006-3568(2006)56[967:OPABM]2.0.CO;2]

Wise, M., Calvin, K., Thomson, A., Clarke, L., Bond-Lamberty, B., Sands, R., Smith, S.J., Janetos, A., Edmonds, J. (2009). Implications of limiting CO2 concentrations for land use and energy. *Science, 324*(5931), 1183-1186.
[http://dx.doi.org/10.1126/science.1168475] [PMID: 19478180]

Zhang, H., John, R., Peng, Z., Yuan, J., Chu, C., Du, G., Zhou, S. (2012). The relationship between species richness and evenness in plant communities along a successional gradient: a study from sub-alpine meadows of the Eastern Qinghai-Tibetan Plateau, China. *PLoS One, 7*(11), e49024.
[http://dx.doi.org/10.1371/journal.pone.0049024] [PMID: 23152845]

CHAPTER 3

Human Ecology: A New Perspective

Abstract: Complexity exists in systems with simple architecture. The unit of architecture, in this context, is a predator–prey community. In case another predator invades the patch in which this community inhabits, temporal dynamics would go chaotic. Chaotic dynamics is characterized by short–term predictability. This leads to **Predator-induced phenotypic plasticity.** It has been found in Daphnia's Neuro - physiological mechanisms of Ad hoc environmental phenotypic adaptation. **Induced defenses** in *Daphnia*; a prey for fish, phantom midge larvae, tadpoles, and several aquatic insects, engage in predation-specific chemical cues that signal increased predation risk. Identification of friends and foes is facilitated by Chemo-receptors in Daphnia. Olfactory receptor (OR) neurons belong to the G – protein-coupled – receptor super family. These neurons get activated when air-borne molecules bind to ORs expressed on their cilia. Transport of goods and services involves the movement of vehicles that release NO_2 into the environment.

Molecular switches on plant leaves help sense their environment. These switches are 10^{-15} m long molecules made of **femto particles**. Animals and 'social animals' (individual humans living in different societies) interact with each other through members of G – protein-coupled receptor superfamily. Animals consume plants that provide food, fodder, fuel, and fiber for the growing human population. **Social Capital**, a network of relationships among people in an efficient society, creates **Human capital**; good health, and knowledge of things, which are useful for the execution of duties of an employee in a company. The chapter provides a crisp description of all that goes into different aspects of human ecology. This discipline puts humans on the center stage. An integration of this discipline with neurosciences would broaden the scope of both disciplines.

Keywords: Anthropocentric, Complex systems, Cultural dimensions, Dynamical complexity, G-protein, Human capital, Hierarchical structures, Neuroscience, Predator-prey community, Predator-induced phenotypic plasticity, Receptor superfamily, Social environments, Social capital, Sustainable development.

INTRODUCTION

Human Ecology is the study of the relationship between humans and their natural, social, and built environments. The human dimension of Ecology is the main focus of this discipline. Herbert Spencer's view of Ecology reminds us of a discipline that enquires into the *patterns and processes* of interaction of humans

Vikas Rai

with their environments. The evolution of structures; both social and natural, is the focalpoint. It recommends the study of humans as living systems in complex environments; social, natural, and built. One of the main objectives of this discipline is ***sustainable development***.

Natural environments embrace all living and non-living things; vegetation, micro-organisms, soil, rocks, atmosphere, and natural events (Clayton, 2003). Forest fires, floods, and cyclones are a few examples of natural environments. The effect of natural events on living things causes drastic changes and sometimes irreversible changes in the dynamics of ecological entities; *e.g.*, forests and lakes. The potential of ecological processes; regeneration and succession of vegetation, decides how fast a forest returns to its original state (Allen and Starr 1982). Forests and Lakes are ***complex systems***. Ant-hills and ants themselves, human beings and their economies, nervous systems, and nerve cell themselves, are a few examples of complex systems with hierarchical structures (Fig. **1**). The economy is yet another example (Anderson, *et al.* 1988; Glass, 2001; Kaufman, 1993; Nicolis and Nicolis, 2007) with hierarchical structures (Lutz, 2000; Li – he & Dong – Sheng, 2004, Sales – Pardo, *et al.* 2007). There could be systems without hierarchical structures. Those without hierarchical structures are difficult to observe and understand (Simon, 1962). The former class of complex systems have multi-level organization and are decomposable.

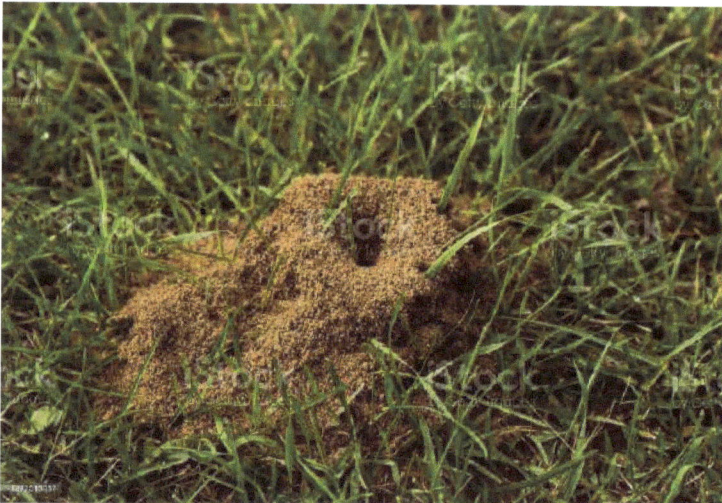

Fig. (1). Ant hill on a meadow in wisconsin state park.

Components (individual units) of complex systems interact nonlinearly; *i.e.*, for such a system rule "*two plus two equal to four*" does not hold. The outcome of the mathematical operation of adding two to two lies between zero to four.

Sometimes, the operation may yield '4'. One should note that adding and multiplying two to two is the same. It implies that there cannot be a way to get the 'whole' by assembling the parts. Complex systems with hierarchical structures exist. A building with several floors; with individual rooms on each floor connected by pathways, is a classic example. A complex system can be decomposed into subsystems (components) with different scales. Thus, **multi–level organization** is its intrinsic property. The mathematical operation of adding 'two' to 'two' would mean that these individual components (or units) have two interacting species; *e.g.*, a predator–prey community distributed in two nearby patches (Fig. **2**). The predator is a specialist; *i.e.*, it has limited food options. Individuals of both species are allowed to cross over and move to the other patch. Predator-prey interaction is nonlinear (Rai, 2013). This is an example of a complex system without a hierarchical structure. The dynamics of this simple system can be complex depending on the value of the parameter, which measures the intensity of migration.

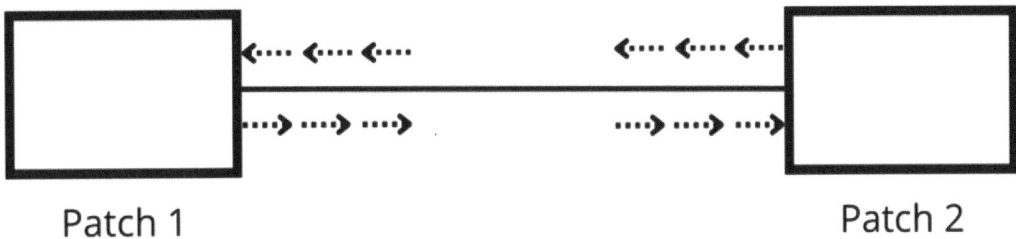

Patch 1   Patch 2

Fig. (2). The emergence of dynamical complexity in a two-patch system linked by migration in a fragmented landscape.

Social Environments refer to the immediate physical and social setting in which people live. It includes a culture that the individual was educated or lives in, people and institutions; marriage and family, beliefs and practices, *etc*. A new model has been developed to explore how changing an individual's certainty in the belief on the truth of one statement leads to changes in their beliefs on the truth of others. This tool has the potential to help estimate likelihood of being persuaded into a new belief. Opinions are rarely formed by accepting or rejecting the social consensus of others.

Social capital refers to the network of groups of individuals having a common goal. Thus, unity of purpose is the connecting thread. The concept of social capital as a resource for action provides an instrument to introduce social structure into the rational action paradigm. Social capital expresses itself in the following forms:

- Obligations and expectations,

- Information channels

- Social norms

Feelings of trust and safety, reciprocity and participation are instrumental in building values, norms, and outlook in life. Productive consumption leads to the development of human capital, which in turn, self-organizes into social capital. Coleman has examined the lack of social capital available to high school sophomores on dropping out of school before graduation. Human capital refers to the economic value of the experience and skills of a worker. Children's environmental identity refers to their interests in and emotional attachment to the natural world. Children recognize their environmental identity but science classrooms are unaware of such recognition (Tugurian & Carrier, 2017).

SOCIAL CAPITAL PARTICIPATES IN THE CREATION OF HUMAN CAPITAL

Social networks and community involvement lead collectively to positive health consequences. It has been observed that those persons who are socially engaged with others and actively involved in their communities have a higher life expectancy and enjoy sound physical and mental health (Leyden, 2003). Social, and community engagement, helping the state to construct roads and parks for the people of all communities in the society without any discrimination, has positive health effects (Walker, 2014).

Built Environments refer to surroundings created by humans, for humans, and to be used for human activities; *e.g.*, buildings, roads, and parks. Buildings create **Indoor Environments.** These are complex in the sense that those living in the building are exposed to a variety of gases and particles from construction activities and carpets (Samet & Spengler, 2003); *e.g.*, ongoing construction work in a building nearby multi-storey building housing corporate offices on different floors. Smith (1976) describes a study of residential neighborhoods and its multi–dimensional character. Planning for the community care of the mentally ill demands knowledge of the parameters of architectural design of residential neighborhoods which guarantee well-being of patients with diseases caused by stress and pollution. Human-induced disturbances adversely affect our food systems. The impact of food and eco-environments on human health, and ecological determinants of health have been discussed by Li (Li, 2017). Efforts to minimize the adverse effects of human intervention on biodiversity should be a priority.

SUSTAINABILITY AND CULTURE

Water is a critical resource. Its role in the lives of humans and other species is well known. Plants cannot survive without water. Dawson (1993) discussed the hydraulic lift of water through the stems of plants and their interactions with other plants in natural environments. We learn about sustainability from ancient texts by Francis of Assisi, Nikolaus von Kues, Spinoza, von Lenné, Goethe, and Novalis. The first ecologists Alexander von Humboldt and Ernst Haeckel, founders of ecology as the science of ecosystems have contributed to our understanding of the relationship between **sustainability** and **culture**. The debate on the *Limits to Growth* after 1972, the Stockholm Conference on the Environment, the reports submitted by the Brandt and Brundtland Commissions, the Gaia Hypothesis, newly coined terms like biodiversity, ecological footprint, renewable resources – and, of course, the 'Spirit of Rio' and the international agreements (the Climate, the Biodiversity, and the Desertification Convention) inspired by it all impress the importance of sustainability. Four decades of this kind of 'earth politics' has led to the realization that the planet Earth is still locked on a collision course. Groeber (2012) submits that there is either not enough differentiation or none at all between 'sustainable' and 'non-sustainable'. The litmus test for all economic and social developments and political decisions is to ask ourselves: 'Does it reduce the ecological footprint?'

and 'Does it widen access to a good quality of life?'

The anthropocentric approach asserts that humans are superior to nature. Plants and animals are resources to be consumed by them. Contrary to this, the **Cosmo-centric approach** advocates the idea that the natural state of our universe should remain unchanged by human habitation irrespective of population pressure. Sustainability is the guiding principle of human consciousness. In this context, the question that is asked is how much one should consume?

CULTURAL DIMENSIONS OF SUSTAINABILITY

Culture is a way of life. Its components are Vidya, Sadhana, and Kala (Aurobindo, 1999). Heritage represents the identity of people (McLean, 2006). Sustainability studies are a norm. UNESCO's World Heritage Convention in 1992 introduced the concept of **Sacred Natural Sites** (SNS) in a debate on sustainability. Relationships between conservation practitioners and indigenous communities define the internal social structure of world society (Rode, 2015). IUCN–UNESCO guidelines for the conservation and management of SNS in 2008, adjudged local people as "custodians" of SNS. Godbout's concept of 'gift beyond debt' comes into the picture at this juncture (2006). SNS are gifts beyond debt. It is the duty of local people to take care of these societal assets.

Belief systems have two basic components: 1) cognitive and 2) motivational. Information represented in a neuronal population code can be represented by a representational distance matrix in a representational space. Analysis of representational geometry helps us compare models and the brain. Brain computation can be regarded as a transformation of representational similarity structure (Kriegeskorte and Kievit, 2003). Algorithmic approaches have been developed to understand how a consensus is reached among individuals or a group of people. Belief system dynamics occur in both small groups and large populations. But, it is more potent for the small group. A belief system is defined as a configuration of ideas and attitudes in which elements are bound together by some form of constraint or functional interdependence (Converse, 1964). It has been demonstrated that the existence of logical constraints facilitates convergence to a shared belief system in a small group of individuals (Birch *et al.*, 2001; Friedkin *et al.*, 2016). Structural elements of small group interpersonal influence systems and rules regulating debates among individuals or groups are to be discerned.

ECOLOGY AND NEUROSCIENCE

Social interactions, *e.g.*, foraging, mating, and defending territories against groups of individuals from other species involve signaling between individuals of groups. It would be interesting to explore cellular networks (Blankenship & Feller, 2010) participating in signaling that cares for group interest ignoring individual interest.

Questions that need to be answered are: *Is there something like social choice?* Which attributes of ecological networks decide social choice? The following pertinent research questions are open.

1. Which sensory pathways dominate social choice?
2. Why fitness of individuals differ?
3. Is there any cross–talk between different pathways? If yes, when and why?

The role of infra–slow oscillations (Gorin *et al.*, 2016; Shadlen & Newson, 1998); i. e., oscillations having frequency less than 0.1 Hz, episodic 'up' and 'down' state in questions 1 to 3 (Buzsaki & Draghan, 2004; Mizuseki, *et al.*, 2009; Rieke, 1997) may be explored. Answers to the aforementioned questions are to be worked out by invoking the following themes.

1. Group interest *vs*. individual interest,
2. Social choice *vs*. individual choice,
3. Survival of the fittest, natural selection, longevity, demographic parameters, genetic inheritance.

M. Gagliano (2015) has discussed the cognitive nature of plants. Plants perceive, assess, learn, and remember. All these happen without cellular brain architecture; memory without brain and neural tissues is something strange in them. Information acquisition, memory, and learning are all part of their daily lives. Cognition is a process, not a property. The vascular systeminteracts with its environment and modifies its behavior in response to varying environments. It is a challenge to work out mechanisms and operative principles in this case. Sensory pathways include the auditory olfactory bulb (AOB), amygdala and hypothalamus regions of the brain. The sensory ecology of *predator–induced phenotypic plasticity* has been reviewed by Weiss (2019). Daphnia's Neuro - physiological mechanisms of Ad hoc environmental phenotypic adaptation are laid bare. The author discusses induced defenses in prey, *Daphnia*. Daphnia are prey for fish, phantom midge larvae, tadpoles, and several aquatic insects. Predation-specific chemical cues generate signals of increased predation risk. Identification of friends and foes is facilitated by Chemo-receptors in Daphnia. Olfactory receptor neurons are bipolar neurons. These receptors belong to the G-protein-coupled receptor superfamily. Neurons get activated when airborne molecules bind to ORs expressed on their cilia.

CONCLUSION

In this chapter, key concepts in human ecology are presented. This discipline puts *homo sapiens* at the center stage. Humans feed on animals that consume plants. **Molecular switches** on plant leaves help sense their environment. These switches consist of 10^{-15} m long molecules made of **femto particles** (Rai 2024). Animals and 'social animals' (individual humans living in different societies) interact with each other through members of the G protein-coupled receptor superfamily. One of the key challenges is to find out the mechanism through which plants sense their changing environment. In the absence of a neural architecture, the entangled pairs of proton and anti–proton are proposed to play a critical role in this sensing. Proton pumps exist in plant cells (Cosse and Seidel, 2021). Plasma membrane ATPase serves as the dominant proton pump. Anti-protons are present in the environment. One of the major concerns of human ecology is to understand how an evolving relationship between nature and culture shapes our world (Steiner, 2016). Therefore, the author feels that the cultural aspect of human ecology should be given proper attention. This is why it is discussed in detail. It advocates for an integration of ecological sciences with those of neurosciences.

Economies run human society. Humans cannot survive without industries. Industries cause pollution; *air, water, noise, etc*. The idea that the waste material of an industry can serve as raw material for another seems promising and needs to be put into practice. This is the theme of the next chapter.

REFERENCES

Allen, T.F.H., Starr, T.B. (1982). *Hierarchy: Perspective for Ecological Complexity.*. Chicago: University of Chicago Press.

Anderson, P.W., Arrow, K.J., Pines, D. (1988). *The economy as an evolving Complex system.*. Redwood City, CA: Addison – Wesley.

Aurobindo, S. (1999). *The Human Cycle:The Psychology of Human Development.*. Twin Lakes, Wisconsin: Lotus Press.

Birch, L.L., Fisher, J.O., Grimm-Thomas, K., Markey, C.N., Sawyer, R., Johnson, S.L. (2001). Confirmatory factor analysis of the Child Feeding Questionnaire: a measure of parental attitudes, beliefs and practices about child feeding and obesity proneness. *Appetite, 36*(3), 201-210. [http://dx.doi.org/10.1006/appe.2001.0398] [PMID: 11358344]

Blankenship, A.G., Feller, M.B. (2010). Mechanisms underlying spontaneous patterned activity in developing neural circuits. *Nat. Rev. Neurosci., 11*(1), 18-29. [http://dx.doi.org/10.1038/nrn2759] [PMID: 19953103]

Buzsáki, G., Draguhn, A. (2004). Neuronal oscillations in cortical networks. *Science, 304*(5679), 1926-1929. [http://dx.doi.org/10.1126/science.1099745] [PMID: 15218136]

Buzsáki, G., Logothetis, N., Singer, W. (2013). Scaling brain size, keeping timing: evolutionary preservation of brain rhythms. *Neuron, 80*(3), 751-764. [http://dx.doi.org/10.1016/j.neuron.2013.10.002] [PMID: 24183025]

C. Li – he and W. Dong - Sheng "Hierarchical Self– organization of Complex Systems,". *Chem. Res. Chin. Univ., 20*(4), 440-445.

Clayton, S.D. (2003). *Identity and the Natural Environment: The Psychological Significance of Nature.*. MIT Press. [http://dx.doi.org/10.7551/mitpress/3644.001.0001]

Converse, P.E. (1964). The Nature of Belief Systems in Mass Publics. In: Apter, D.E., (Ed.), *Ideology and Discontent.* (pp. 206-261). New York: The Free Press.

Cosse, M., Seidel, T. (2021). Plant Proton Pumps and Cytosolic pH-Homeostasis. *Front. Plant Sci., 12*, 672873. [http://dx.doi.org/10.3389/fpls.2021.672873] [PMID: 34177988]

Dawson, T.E. (1993). Hydrolic lift and water use in plants: implications for performance, water balance and plant - plant interactions. *Oecologia, 95*(4), 565-574. [http://dx.doi.org/10.1007/BF00317442] [PMID: 28313298]

Friedkin, N.E., Proskurnikov, A.V. (2016). R. tempo and R. Parsegov, "Network science on belief system dynamics under logic constraint,". *Science, 354*(6310), 321-326. [http://dx.doi.org/10.1126/science.aag2624] [PMID: 27846564]

Gagliano, M. (2014). In a green frame of mind: perspectives on the behavioural ecology and cognitive nature of plants. *AoB Plants, 7*, 7. [http://dx.doi.org/10.1093/aobpla/plu075] [PMID: 25416727]

Glass, L. (2001). Synchronization and rhythmic processes in physiology. *Nature, 410*(6825), 277-284. [http://dx.doi.org/10.1038/35065745] [PMID: 11258383]

Godbout, J.T. (2006). Le don au-delà de la dette. *Rev. MAUSS, n o 27*(1), 91-104. [http://dx.doi.org/10.3917/rdm.027.0091]

Gorin, M., Tsitoura, C., Kahan, A., Watznauer, K., Drose, D.R., Arts, M., Mathar, R., O'Connor, S., Hanganu-Opatz, I.L., Ben-Shaul, Y., Spehr, M. (2016). Interdependent conductances drive infra - slow intrinsic rhythmo - genesis in a subset of Accessory olfactory bulb projection neurons. *J. Neurosci., 36*(11), 3127-3144.

[http://dx.doi.org/10.1523/JNEUROSCI.2520-15.2016] [PMID: 26985025]

Groeber, U. (2012). *Sustainability. A cultural history* Translated from the German by Ray Cunningham, Totnes, UK: Green Books. ISBN: 978-0-8578

Kaufman, S. (1993). *The Origins of Order..* New York: Oxford University Press.
[http://dx.doi.org/10.1093/oso/9780195079517.001.0001]

Kriegeskorte, N., Kievit, R.A. (2013). Representational geometry: integrating cognition, computation, and the brain. *Trends Cogn. Sci., 17*(8), 401-412.
[http://dx.doi.org/10.1016/j.tics.2013.06.007] [PMID: 23876494]

Leyden, K.M. (2003). Social capital and the built environment: the importance of walkable neighborhoods. *Am. J. Public Health, 93*(9), 1546-1551.
[http://dx.doi.org/10.2105/AJPH.93.9.1546] [PMID: 12948978]

Li, A.M.L. (2017). Ecological determinants of health: food and environment on human health. *Environ. Sci. Pollut. Res. Int., 24*(10), 9002-9015.
[http://dx.doi.org/10.1007/s11356-015-5707-9] [PMID: 26552789]

McLean, F. (2006). Introduction: Heritage and Identity. *Int. J. Herit. Stud., 12*(1), 3-7.
[http://dx.doi.org/10.1080/13527250500384431]

Mizuseki, K., Sirota, A., Pastalkova, E., Buzsáki, G. (2009). Theta oscillations provide temporal windows for local circuit computation in the entorhinal-hippocampal loop. *Neuron, 64*(2), 267-280.
[PMID: 19874793]

Nicolis, G., Nicolis, C. (2007). *Foundation of Complex Systems..* Singapore: World Scientific.
[http://dx.doi.org/10.1142/6253]

Odum, E. P. (2005). *Fundamentals of Ecology*Thompson brooks/Kole.

Rai, V. (2024). *The Brain: A Systems Neuroscience Perspective..* Sharjah, UAE: Bentham Science Publishers.
[http://dx.doi.org/10.2174/97898152569871240101]

Rieke, F., de Ruyter van Steveninck, R., Bialek, W. (1997). *Spikes: exploring the Neural code..* Cambridge, MA: MIT Press.

Rode, R. (2015). *Recognition of indigenous people's sacred natural sites and the cultural dimension of sustainable development, Perceptions of sustainability in heritage study, edited by Marie – Theres Albert.* (pp. 125-136). Berlin: De Gruyter.

Sales-Pardo, M., Guimerà, R., Moreira, A.A., Amaral, L.A.N. (2007). Extracting the hierarchical organization of complex systems. *Proc. Natl. Acad. Sci. USA, 104*(39), 15224-15229.
[http://dx.doi.org/10.1073/pnas.0703740104] [PMID: 17881571]

Samet, J.M., Spengler, J.D. (2003). Indoor environments and health: moving into the 21st century. *Am. J. Public Health, 93*(9), 1489-1493.
[http://dx.doi.org/10.2105/AJPH.93.9.1489] [PMID: 12948968]

Schnore, L.F. (1961). Geography and Human Ecology. *Econ. Geogr., 37*(3), 207-217.
[http://dx.doi.org/10.2307/142087]

Shadlen, M.N., Newsome, W.T. (1998). The variable discharge of cortical neurons: implications for connectivity, computation, and information coding. *J. Neurosci., 18*(10), 3870-3896.
[PMID: 9570816]

Simon, H.A. (1962). The architecture of complexity. *Proc. Am. Philos. Soc., 106*, 467-482.

Smith, C.J. (1976). Residential neighborhoods as humane environments. *Environ. Plann. A, 8*(3), 311-326.
[http://dx.doi.org/10.1068/a080311]

Steiner, F. (2016). *Human Ecology: How Nature and Culture Shape Our World.Island Press, February.*

Stowers, L., Logan, D.W. (2010). Sexual dimorphism in olfactory signaling. *Curr. Opin. Neurobiol., 20*(6),

770-775.
[http://dx.doi.org/10.1016/j.conb.2010.08.015] [PMID: 20833534]

Tugurian, L.P., Carrier, S.J. (2017). Children's environmental identity and the elementary science classroom. *J. Environ. Educ., 48*(3), 143-153.
[http://dx.doi.org/10.1080/00958964.2016.1191415]

Weiss, L.C. (2019). Sensory ecology of Predator – induced phenotypic plasticity. *Front. Behav. Neurosci., 12*, 330.
[http://dx.doi.org/10.3389/fnbeh.2018.00330] [PMID: 30713490]

CHAPTER 4

Elements of Industrial Ecology

Abstract: The central theme of industrial ecology is the idea that the waste material from one industry can serve as raw material for the other. It is **industrial symbiosis**. This eases tension created by the pressure of ever-accumulating industrial waste. Another aspect of an industry is the input of energy. It must be clean; emissions to the environment are minimal, and green, *i.e.*, we must utilize the least amount of resources. The search for clean energy sources has converged to **Hydrogen Energy**. The production efficiency of the electrolysis process is enhanced by the application of an external magnetic field. Solar to hydrogen conversion efficiency attains the levels of economic feasibility with the use of semiconductor sheets made of Rh – co-doped $SrTiO_3$ powders embedded into a gold layer. Solar to hydrogen conversion efficiency (1.1% and 30% quantum yield at 419 nm) was achieved by splitting pure water (pH 6.88). Effects of electric power, external magnetic field, and temperature on conversion efficiency have been investigated and found to be appreciable.

Bio–diesel is another potential source of energy. The energy of sunlight is converted into chemical energy through a biochemical process. A positive aspect of algal cultivation is that it can be grown either in freshwater or brackish water. In this way, it does not compete for fresh water. Microalgae respond by producing more carbohydrates or lipids in conditions of environmental stress; *e.g.*, when a particular nutrient is lacking. Biodiesel derived from algal lipids is ***non-toxic*** and **biodegradable**. Microalgae produce oils 15 – 300 times more than traditional crops per unit cultivated area. The ability of an endolithic cyanobacterial strain, *Leptolyngbya* sp. ISTCY101 to produce biomass from which biodiesel can be produced, has been assessed in experiments conducted exploiting principles of carbon assimilation in the natural ecological niche of the cyanobacterium. Measurements of relevant variables and parameters showed that this strain is capable of returning a reasonably high yield of biomass productivity.

Integration of industrial ecology and ecological economics would expand the scope of the **circular economy**. New information generated in the process of integration would be used to develop new tools for decision-makers. **Self-guidance,** a special attribute, would be available to industry leaders if a course based on this chapter is introduced in the curriculum of engineering undergraduates.

Keywords: Ammonium-oxidizing bacteria, Biochemical process, Cyanobacterial, Endolithic strain, Electrolysis, Hydrogen production, Hydrogen energy, Lipid, Nitrite oxidizing bacteria, Nitrogen fixation, Particulate semiconductor sheets, Sustainable agriculture, Symbiosis.

INTRODUCTION

In the formative stage of **Industrial Ecology (IE)** as a discipline in the 1990s, when it embraced **Social Ecology, IE** was defined by, 1) industrial metabolism, 2) socio-economic metabolism, and 3) urban metabolism. The latter concerns with urban infrastructure frameworks, transportation, waste, energy, greenhouse gas emissions, and food in cities (Kennedy, 2016). Robert Ayres coined the term *'industrial 'metabolism'* in analogy with **biological metabolism**. It refers to the integrated collection of physical processes that convert raw material into energy (Ayres, 1994). It also includes labor input, finished products, and wastes as output. A few of these aspects are discussed in this chapter. A dynamic industrial development with structural changes, permitting investments in modern technology and emission control with waste minimization, environmental auditing, political and industrial decision–making, human behavior, and attitudes, creates a livable society (Anderberg, 1998).

Industrial Ecology is a sub-discipline of ecological science that is concerned with the optimum utilization of natural resources and minimization of damage to the environment caused by human activities (*e.g.*, setting up industries and developing the real estate, *etc.*). Frosch (1992) put the subject of "industrial ecosystems" in the mainstream. What emerged in the nineteenth century is now known as *Industrial Symbiosis*. It is an association between two or more industrial facilities or companies in which the wastes or byproducts of one is used as raw material by the other. Thus, the cost of waste disposal is reduced. Residues and byproducts provide new avenues. Wastes do not go to the landfill, carbon emissions are contained and new business opportunities are generated in the process. The natural world favors *industrial synergy* as it is advantageous for both the parties involved and leads to both commercial and environmental advantages. A sequence of independent economically driven actions led to the establishment of an industrial district at Kalundborg, Denmark. The functioning of other industrial districts was analyzed in the light of the Kalundborg Industrial Ecosystem, which is essentially an eco-industrial park. Two closely located industrial units in symbiosis (Ehrenfeld & Gertler, 1997) are shown (Fig. **1**). Flow of both energy and matter is shown.

It is an industrial network with partnerships between private and public companies located in the area. Kalundborg eco-industrial park, a classic example of industrial symbiosis, saves 24 million Great Britain Pounds annually. It reduces CO_2 emission to the atmosphere by 635,000 tons, 3.6 million cubic meters of water, and 100 GHz of energy. In addition to these savings, it contributes 87000 tons of solid materials. This sets a benchmark for the evaluation and analysis of other similar projects.

Fig. (1). The Kalundborg Industrial Ecosystem; an Eco-industrial park based on a circulation model of production. The basic principle behind its operation is: that matter is convertible into energy. The figure shows the flow of materials and energy. Source: University of Southern California, School of Engineering's blog.

Industrial Ecology is a study of the interaction between industry and the environment. It is based on the circular model of production; the waste of an industry serves as a raw material for the other industry (Graedel & Allenby, 1998). The theoretical and practical framework of the circular economy needs to be expanded to help sustainability. Analysis of the flow of matter and energy was carried out using concepts and tools borrowed from other fields. Bruel *et al.* (2018) recommend the reconciliation of IE and EE in order to develop better tools that help in decision-making. A study of socio-economic consequences of these flows would generate new information. This new information along with consideration of Earth's carrying capacity, and behavioral and social aspects of the natural system, would lead to a model of socioeconomic system. Analysis of this model system would guide us to sustainability (Bruel *et al.* 2018).

Damage to the environment is minimized if clean fuel is used in motor vehicles. Human society has been looking for clean energy sources; sources not polluting the environment. Biodiesel is a clean fuel. In the next section, we present different aspects of biodiesel production.

BIODIESEL AS A CLEAN FUEL

Biodiesel is produced from vegetable oil/animal oil/ fats, tallow, and waste cooking oil through *trans- esterification*; a chemical reaction in which alcohol (generally methanol) exchanges its R group with the R' group of a carboxylic acid ester. Potential sources of these oils are rapeseed, palm, or soybean. It is environmentally friendly as it produces no net carbon in the form of CO_2. The amount of CO_2 absorbed during the growth of oil plants is reduced in the combustion process. Biodiesel can be produced through any of these routes:

1. Base-catalyzed Trans - esterification of the oil,
2. Direct acid-catalyzed Trans - esterification of the oil, and
3. Conversion of the oil to its fatty acids and then to biodiesel.

The most common process producing biodiesel is base catalyzed Trans - esterification as it requires low temperature and pressure. The efficiency of the process is 98%. Alternatives to first generation biofuels are those of second and third-generation. The former is derived from lingo-cellulosic biomass. Third-generation bio-fuels are produced from algae. The life of algae is critically dependent on 1) sunlight, 2) carbon dioxide, and 3) water. The energy of sunlight is converted into chemical energy through a biochemical process. A positive aspect of algal cultivation is that it can be grown either in freshwater or brackish water. In this way, it does not compete for a so scarce resource, such as fresh water. Microalgae respond by producing more carbohydrates or lipids in conditions of environmental stress; *e.g.*, when a particular nutrient is lacking. Biodiesel derived from algal lipids is ***non-toxic***. It is biodegradable as well. Microalgae produce oils 15-300 times more than traditional crops per unit cultivated area. Singh and Thakur (2015) performed a study to assess the ability of an endolithic cyanobacterial strain, *Leptolyngbya* sp. ISTCY101 to produce biomass from which biodiesel can be produced.

Experiments were designed and conducted by these researchers. Measurements of relevant variables and parameters suggested that this strain is capable of returning a reasonably high yield of biomass productivity.

Glycerol is produced along with biodiesel. After the reaction is complete, glycerol is separated from the methyl ester. The following are benefits that accrue from the process of transesterification.

1. Lowered viscosity,
2. complete removal of glycerides,
3. lowered boiling point,

4. lowered flash point,
5. lowered pore point.

BIOLOGICAL NITROGEN FIXATION AND SUSTAINABLE AGRICULTURE

Nitrogen is an essential plant nutrient. Agricultural systems can be sustainable if the supply of nitrogen to crops exploits this mechanism. It has the following dimensions.

Environmental

The global nitrogen cycle is adversely affected by nitrogen fertilizers. These fertilizers make additions to the existing nitrous oxide in the atmosphere.

Energy

The primary energy sources for the production of nitrogen fertilizers are natural gas petroleum, and coal. Renewable sources; *e.g.*, plant-synthesized carbohydrates meet the energy requirement in biological nitrogen fixation.

Sustainability

Biological nitrogen fixation provides an ecologically sound mechanism to reuse external nitrogen input and improve the quality and quantity of internal resources at the same time.

Nutrition

It is estimated that approximately 20% of the global food protein is derived from legumes. Nitrification in soils is divided into two categories (Pajares & Bohannan, 2016)

1. Autotrophic, carried out mainly by chemoautotrophic
   a. **Ammonium–oxidizing bacteria**; *e.g.*, *Nitrosomonas*, *Nitrosospira*, and *Nitrosococcus* species and archaea; *e.g.*, *Nitrososphaera*, and *Nitrosotalea* lineages from the phylum Thaumarchaeota), and
   b. **Nitrite - oxidizing bacteria**; Nitrobacter, Nitrospira, and Nitrococcus species
2. Heterotrophic nitrification is carried out by heterotrophic bacteria and fungi with the potential to oxidize both organic and inorganic compounds like N compounds (Hayatsu *et al.*, 2010).

CLEAN ENERGY SOURCES

Weisz *et al.* (2017) have discovered water channels in Photosystem II. The active site of this complex is a cluster of manganese, calcium, and oxygen ions buried deep within. Using high-resolution mass spectroscopy, these researchers discovered several pathways, which can be used to deliver water to the active site. An understanding of complex mechanisms associated with processes responsible for the ecology and evolution of photo-system II is indispensable for those involved in green energy research. Any renewable source of energy; *e.g.*, solar, wind, *etc.* is considered to be '**green**' if the production of electricity through these sources does not pollute the atmosphere; emissions of greenhouse gases (carbon dioxide, oxides of nitrogen, *etc.*) are minimal.

One of the mandates of industrial ecology is to educate the industry to prepare itself for **self-guidance** so that it can make appropriate choices of sources of energy. Hydrogen energy is such a form of energy. In what follows a few aspects of hydrogen energy are discussed briefly.

HYDROGEN ENERGY: BREAKING FROM THE FOSSIL FUELS

NASA astronauts have been using electrolysis of pure water to supply electricity for the functioning of internal components of space vehicles carrying them since 1972. Research in the production, storage and transport of hydrogen energy picked up the pace .Hydrogen would serve as a source of energy in the future only when produced using one of the renewable energy sources; *e.g.*, **sun, wind, or geothermal energy**. Splitting of water into its components through the process of electrolysis has been the main route. The development of catalysts to enhance the efficiency of hydrogen production through electrolysis has consumed an appreciable amount of time for researchers. Progress made so far in this direction has been reviewed by Wang, *et al.* (2021).

Amorphous to crystalline transitions in Holmium tri - aluminide (Yamamoto, *et al.* 2022) take place when it is heated by an electron beam. An intermediate phase would serve as a better storage. X-ray diffraction and TEM studies have shown that the amorphous phase grows at the expense of the crystalline phase when hydrogen is absorbed by these meta-stable alloys. Wang *et al.* (2016) have demonstrated the scalability and economic feasibility of solar to hydrogen conversion technology using particulate semiconductor sheets, which were obtained by La – and Rh – co-doped $SrTiO_3$ powders embedded into a gold layer. Solar to hydrogen conversion efficiency (1.1% and 30% quantum yield at 419 nm) was achieved by splitting pure water (pH 6.88). Effects of electric power, external magnetic field and temperature on conversion efficiency have been studied (Kartaza *et al.*, 2021, Purnami, *et al.* 2020).

A favorable aspect of the uses of hydrogen energy is that no regulatory authority will dictate its terms to end users as the supply of hydrogen stems from a variety of energy sources (Acar & Dincer, 2018).

Key Points

1. Renewable Energy creates it.
2. It is versatile across industries.
3. It reduces CO_2 emissions.
4. Biomass (steam – methane reforming).

GREEN HYDROGEN

Green hydrogen reduces emissions and cares for our planet. Following domestic resources can be used to produce hydrogen.

1. Water
2. Solar
3. Wind
4. Biomass (steam – methane reforming)

A pertinent question to ask: Is all fuel green? Clean hydrogen is green hydrogen. It produces only water vapor emissions. At present, most hydrogen is produced by fossil fuels. Increased production of fuel-cell electrical vehicles would help contain air pollution. Fuel cells are battery-like entities which do not need re-charging. The USA has planned to have one million cars on the roads of California by 2030.

BLUE HYDROGEN

Why is it called blue? Is it color? No. It is hydrogen produced by natural gas using steam reforming. It brings natural gas and heated water together in the form of steam. High-purity hydrogen is produced from biogas at a biogas plant using a new chemical looping process. The key issue is nationwide availability. A related concept is the **Hydrogen Economy.** It refers to a network of chemical processes that produce, and convert the stored hydrogen to electrical energy at the point of use.

IE tools and practices are as follows:

1. Process Analysis
2. Motion Analysis
3. Time Study
4. Operation Analysis

5. Combination Work Analysis
6. Line Balancing Analysis
7. Material handling

Process analysis refers to the analysis of the production process from raw material to final product in order to minimize waste. IE tools and practices, if implemented properly, help us face the challenges received from global sustainability. The principle of responsible production and consumption (**SDG 12**) must be honored.

CONCLUSION

Thus, the chapter presents a brief introduction to the subject of industrial ecology. *Symbiosis* and *synergy* are its basic tenets. It must be included in the curriculum of a general course in ecology at the undergraduate level in order to prepare industry leaders who possess **self–guidance capabilities**. Linkages between waste disposal, sustainable agriculture, industrial production, and clean energy sources are provided. Hydrogen energy is the most promising among the **clean energy sources**. The present chapter does not provide details of the logistics of hydrogen energy as it can be easily found in other texts; instead, it touches most critical ones; **conversion efficiency of the electrolytic process** and that no legal authority will dictate its terms to end users as the supply of hydrogen stems from a variety of energy sources. It would be challenging to control the cost of hydrogen production by making an appropriate choice of a clean energy source. Another challenge is to keep the delivery of the hydrogen energy safe and sound.

Readers are taken to Chapter 5: *Sustainable Development Goals*. The unifying theme of the present text is that all problems on earth are caused by ecological imbalance. Integration of disciplines; industrial ecology, with ecological economics, requires that Earth's carrying capacity and human behavior and attitudes are considered in the resulting socio–economic system. Analysis of such a system produces new information and equips the decision maker with a new promising tool.

ACKNOWLEDGEMENTS

I. S. Thakur is thanked for making the author aware of his laboratory work related to biodiesel production from *Leptolyngbya* sp. ISTCY101.

REFERENCES

Acar, C., Dincer, I. (2018). Hydrogen Energy Conversion Systems, "Comprehensive Energy Systems,". *Comprehensive Energy Systems, 4*, 947-994.
[http://dx.doi.org/10.1016/B978-0-12-809597-3.00441-7]

Anderberg, S. (1998). Industrial Metabolism and the linkages between economics, ethics and the environment *Ecol. Econ., 24*(2 – 3), 311-320.

Ayres, R.U. (1994). Industrial metabolism: theory and policy, R. U. Ayres and U. K. Simonis (editors), *Industrial Metabolism: Restructuring for Sustainable Development* United Nations University Press: Tokyo, Japan.

Bruel, A., Kronenorgy, J., Trousier, N. (2018). Linking Industrial Ecology and Ecological Economics: A Theoretical and Practical foundation for the circular Economy. *J. Ind. Ecol.,* 1-10. [http://dx.doi.org/10.1111/jiec12745]

Ehrenfeld, J., Gertler, N. (1997). Industrial Ecology in Practice: The Evolution of Interdependence at Kalundborg. *J. Ind. Ecol., 1*(1), 67-79. [http://dx.doi.org/10.1162/jiec.1997.1.1.67]

Frosch, R.A. (1992). Industrial ecology: a philosophical introduction. *Proc. Natl. Acad. Sci. USA, 89*(3), 800-803. [http://dx.doi.org/10.1073/pnas.89.3.800] [PMID: 11607255]

Graedel, T.E., Allenby, B.R. (2009). *Industrial Ecology and Sustainable Engineering..* Pearson.

Hayatsu, M., Tago, K., Saito, M. (2008). Various players in the nitrogen cycle: Diversity and functions of the microorganisms involved in nitrification and denitrification. *Soil Sci. Plant Nutr., 54*(1), 33-45. [http://dx.doi.org/10.1111/j.1747-0765.2007.00195.x]

Kennedy, C.A. (2016). *Taking Stock of Industrial Ecology..* Cham: Springer.

Pajares, S., Bohannan, B.J.M. (2016). Ecology of Nitrogen Fixing, Nitrifying, and Denitrifying Microorganisms in Tropical Forest Soils. *Front. Microbiol., 7*, 1045. [http://dx.doi.org/10.3389/fmicb.2016.01045] [PMID: 27468277]

Purnami, N., Hamidi, N., Sasongko, M.N., Widhiyanuriyawan, D., Wardana, I.N.G. (2020). Strengthening external magnetic fields with activated carbon graphene for increasing hydrogen production in water electrolysis. *Int. J. Hydrogen Energy, 45*(38), 19370-19380. [http://dx.doi.org/10.1016/j.ijhydene.2020.05.148]

Singh, J., Thakur, I.S. (2015). Evaluation of cyanobacterial endolith Leptolyngbya sp. ISTCY101, for integrated wastewater treatment and biodiesel production: A toxicological perspective. *Algal Res., 11*, 294-303. [http://dx.doi.org/10.1016/j.algal.2015.07.010]

Stefan, A. (1998). Industrial metabolism and the linkages between economics, ethics and the environment *Ecol. Econ., 24*(2 – 3), 311-320.

Wang, Q., Hisatomi, T., Jia, Q., Tokudome, H., Zhong, M., Wang, C., Pan, Z., Takata, T., Nakabayashi, M., Shibata, N., Li, Y., Sharp, I.D., Kudo, A., Yamada, T., Domen, K. (2016). Scalable water splitting on particulate photocatalyst sheets with a solar-to-hydrogen energy conversion efficiency exceeding 1%. *Nat. Mater., 15*(6), 611-615. [http://dx.doi.org/10.1038/nmat4589] [PMID: 26950596]

Wang, Z., Zhang, X., Rezazadeh, A. (2021). Hydrogen fuel and electricity generation from a new hybrid energy system based on wind and solar energies and alkaline fuel cell. *Energy Rep., 7*, 2594-2604. [http://dx.doi.org/10.1016/j.egyr.2021.04.060]

Weisz, D.A., Liu, H., Zhang, H., Thangapandian, S., Tajkhorshid, E., Gross, M.L., Pakrasi, H.B. (2017). Mass spectrometry-based cross-linking study shows that the Psb28 protein binds to cytochrome b_{559} in Photosystem II. *Proc. Natl. Acad. Sci. USA, 114*(9), 2224-2229. [http://dx.doi.org/10.1073/pnas.1620360114] [PMID: 28193857]

Yamamoto, T.D., Baptista de Castro, P., Terashima, K., Saito, A.T., Takeya, H., Takano, Y. (2022). Magnetic, thermal, and magnetocaloric properties of the holmium trialuminide HoAl3 with polytypic phases. *J. Magn. Magn. Mater.,* 562169801. [http://dx.doi.org/10.1016/j.jmmm.2022.169801]

CHAPTER 5

Sustainable Development Goals: Good and Bad

Abstract: The central theme of human ecology is sustainable development. United Nations Organization (UNO) in 2015 identified 17 goals; known as *sustainable development* goals (SDGs), to be achieved by 2030. SDG1 (No poverty) and SDG2 (Zero hunger) are difficult to achieve. For the former, a workable measure of poverty is to be evolved. The poverty line defined by the United Nations Department of Economic and Social Affairs (UNDESA) is linked with the Gross Domestic Product (GDP), which varies significantly for countries rich and poor. There is no relationship between poverty in the USA and India. It is *relative poverty*. A universal absolute poverty, which is not linked with Gross Domestic Product, needs to be considered. The prospect of whether the goal of zero hunger would be achieved, depends on the state of **sustainable agriculture** in a country at any given time. SDG2 may be achieved by 2030 with cooperation among rich and poor countries. If developing countries are provided soft credit by developed countries from time to time, a few targets could be achieved.

Agricultural productivity depends on capital. The interaction of disease and human capital leads to dynamics in the state space of the system represented by multiple equilibria (two stable equilibria and an intervening unstable equilibrium). If compared with the famous Lorenz attractor, which presents trajectories of two convective cells; one lying over the other, in the state space of the system (the bottom convective cell is heated up from below, with two unstable foci and an intervening saddle point), it is clear that the interaction of disease and human capital would generate oscillations in system's state space. This explains why agricultural productivity varies; and oscillates between two states of low and high productivity. The incidence of unpredictable epidemics in this system would lead to chaos; which allows only short-term predictability. Therefore, SDG 3 (Good Health and Well–Being) appears to be wishful thinking. This knowledge adds value to SDG 12 (**responsible consumption and production**). Production refers to both agricultural and industrial.

Occupational Choice (SDG 3, SDG 8) is a critical factor. It depends on the beliefs and practices of the people of a nation. Banerjee & Newman (1993) developed a model of economic development. Economic development is considered as a process of **institutional transformation.** Capital market imperfections drive the dynamics of the system considered. Depending on the initial distribution of wealth, the economy generates two scenarios:1) either widespread cottage industry or factory production, 2) prosperity or stagnation. An individual's decision of occupation depends on whether he/she is wealthy or poor. The poor go for employment contracts (factory production) and the wealthy go for entrepreneurship (widespread cottage industry). A society needs both kinds of people. An economy that is poised between two scenarios is desirable.

SDGs 13, 14, 15, and 17 are linked with each other in the sense that rain depends on tree cover present on the land surface. Water bodies receive water through precipitation which depends on the interaction between the sun and ocean; the reservoir of resources. Forest cover and land use patterns also affect climate. If rich countries help poor countries under the aegis of UNO through its different developmental programs, a few of the SDGs can be partially achieved. If developed nations continue to exploit situations in poor (developing) countries, then, there is no hope.

Keywords: Climate change, Distributive justice, Freedom, Family planning, Gross national product, Gross national income, Homeostasis, Industry, Institutional transformation, Occupational choice, Occupational safety, Poverty, Population growth, Sustainable agriculture, Sustainable human settlements.

INTRODUCTION

A gap between two perspectives (*i.e.*, between an exclusive concentration on economic wealth and a broader focus on the lives we can lead) has been identified as a central theme (Sen, 1997) of human life. In his opinion, democracy and human rights hold the key to **sustainable development** with good development goals. Human life faces many challenges in its journey; *e.g.*, epidemic outbreaks (DON, 2023), pandemic waves (Madhav, *et al.* 2017, Rai & Sarita, 2023), and Droughts and famines (Sen, 1997). Droughts and famines are not unfamiliar events. The welfare economist argues that the famine presents a situation in which people do not have sufficient buying power. It is not the shortage of food grains.

The reader's attention is drawn to a few examples of both.

Droughts and Famines

It is an extended period of extremely dry weather when there is not enough rain to support agricultural activities; *e.g.*, sowing paddy crops for food grain production and consumption. It could also cause a shortage of drinking water when the period of dry weather gets elongated. A linked event is famine. A few examples are given below.

Africa

1983 - 1985 Ethiopia famine

2011 East Africa Drought

India

The Great Bengal famine of 1770 struck during the period (1769 – 1770). It affected 30 million people. Bihar famine (1966 – 1967), 2353 deaths were reported by the state. It was happily tackled by the state.

Bangladesh

After independence in 1971, Bangladesh confronted a famine in 1974.

Developing Countries are located in the **Global South**; a geopolitical group of poor or middle–income states. It is located in a broad strip ranging from South-East Asia, Africa, and Pacific islands, all the way to the Caribbean to Latin America. The history of Europe suggests that two countries, France and Britain, were imperialists; *i.e.*, expansion of territory by way of invasion was their agenda. Common Wealth Nations are a group of countries ruled by Britain; North American countries, Canada, and Australia, are examples. These are in different subcontinents. In Canada, Wars were fought and peace was brokered between two Britain and France by a peaceful settlement.

Two major reasons for World War II were: 1) the worldwide **economic depression** and 2) the failure of the **League of Nations.** In order to have an idea of the magnitude of the Holocaust, World War II (1941 – 1945), it is suffice to note that Nazi Germany and its allies murdered nearly six million Jews across Europe-occupied Germany. The world economy was devastated by WWII as the war was long drawn. In an attempt to establish international peace and ensure global security, the United Nations Organization was established. It began its operations on October 24, 1945.

Poverty and other deprivations go hand-in-hand with health and educational concerns, and to reduce inequality, economic growth needs to be enhanced to ensure the preservation of nature. United Nations Conference on 'Sustainable Development' **took the initiative to start deliberations among member states to evolve goals that guarantee** sustainable development. A chronology of events is given below.

Chronology of Events

1992 In Earth Summit, Rio de Janeiro Brazil, more than 178 countries adopted Agenda 21, Global Partnership to improve quality of life, which is dependent on how little damage we inflict on the environment

2000 Millennium Summit emphasized Sustainable patterns of consumption, production, economic and social development

2002 The Johannesburg Declaration on Sustainable Development and the plan of implementation adopted at the World Summit on Sustainable Development in South Africa, reaffirmed the global community's commitment to poverty eradication.

2012 Outcome document "The Future We Want" was brought out in Rio de Janeiro, Brazil.

2013 General Assembly set up a 30-member Open Working Group to develop a proposal on SDGs.

2015a General Assembly began what is called the Negotiation Process.

2015b Addis Ababa Action Agenda

ADDIS ABABA ACTION AGENDA: FINANCING SUSTAINABLE DEVELOPMENTS

A framework to align flows of finance and policies with economic, social, and environmental priorities was evolved by reaching a consensus by heads of state and of Governments, and representatives gathered in Addis Ababa from (13 - 16) July 2015 to address the challenges of financing and creation of an environment which empowers member nations to achieve sustainable development at all levels through global partnership and solidarity (TICFD, 2015). It is a holistic and forward-looking framework to follow up on commitments and assess the progress made in the implementation of the Monterrey Consensus and Doha Declaration.

The world made significant progress in economic activities as flows of finance were smoothed after the Consensus was reached. The following are the main items on the agenda.

1. Peaceful and inclusive society with equitable global economic system
2. Reinvigorate the financing for development
3. To reduce the number of people in extreme poverty
4. To ensure that development efforts enhance resilience in the face of natural disasters, disease outbreaks, environmental degradation and climate change, and
5. To ensure efficient and transparent mobilization and use of resources in light of the principles laid down in the Rio Declaration on Environment and Development (UNDESA, 2023).

In the next section, SDGs are discussed in detail. United Nations member states adopted a resolution with seventeen development goals in 2015 to be achieved by 2030.

SUSTAINABLE DEVELOPMENT GOALS

SDGs

1. No poverty

2. Zero Hunger

3. Good Health and Well – being

4. Quality Education

5. Gender Equality

6. Clean Water and Sanitation

7. Affordable and Clean Energy

8. Decent Work and Economic Growth

9. Industry, Innovation and Communities

10. Reduced Inequalities

11. Sustainable Cities and Communities

12. Responsible Consumption and Production

13. Climate Action

14. Life Below Water

15. Life on Land

16. Peace, Justice Strong Institutions, and

17. Partnerships for the Goals

Now, the chapter presents a brief description of each one of them.

SDG 1 No Poverty

Definition

World Bank Organization defines poverty by the statement: ***poverty is hunger***, ***poverty is lack of shelter***. Poverty is more than the lack of 'income and resources' (Sen, 1997). It also includes economic facilities and social opportunities.

The poverty line was set at 1.90 USD per day initially. It has been revised as the prices of food grains and the cost of living rose. An individual, who earns less than 2. 15 USD per day is poor. This definition of poverty is not fair. The poverty line in the US, which is linked with the US median income, has no relation with the poverty line in India. It is only the relative poverty that is reset from time to time. Abhijit Banerjee (Banerjee &Duflo, 2019) recommends considering a universal absolute poverty line that is not linked with the gross domestic product (GDP) of any specific country. A network of 181 Professors at universities worldwide, The Abdul Latif Jameel Poverty Action Laboratory (J – PAL) works on the 'eradication of poverty'. J -- PAL assumes that policy is judgment informed by scientific action. Official development assistance (ODA) is the grant from all donors that focus on poverty reduction as a share of the recipient countries gross national income (gni). Gross national income is the total amount of money earned by people of a nation and businesses inside and outside the country (UNDESA, 2023).

There are seven targets. Only two of them are mentioned here.

Target 1.1 Eradicate Extreme Poverty

Economist Abhijit Banerjee (Director, J – PAL laboratory, MIT, USA) has suggested to the Prime Minister of India to provide Rs. 6000 per month to every Indian in extreme poverty.

Target 1.2 Reduce poverty at least by 50 percent.

Universal Absolute Poverty can be used to achieve this goal. It is being tested by economists by carrying out field trials.

Now, a natural question to ask is: Why a textbook in ecology should discuss poverty? Poverty compels poor people in rural areas to confine themselves to marginal lands. This leads to an increase in ecological vulnerability, landslides, *etc*. Poor neighborhoods are marked by bad management practices of waste collection and disposal. High rates of mortality due to parasitic and infectious diseases are a hallmark of these neighborhoods. Measures to stop deforestation and encouragement for re-forestation should be put into practice to create a sustainable resource base for the poor. Models of persistent poverty designed by economists do not take into account complex biophysical processes into account. Ngonghala, *et al*.(2014) explored the relationship between poverty, disease, and ecology in the framework of complex systems. The authors considered a coupled model system of infectious diseases and economic growth. ***Poverty traps*** emerge from nonlinear relationships determined by a number of pathogens in the system.

An analysis of the model suggests that the generation and maintenance of poverty traps can be understood by combining insights from both **ecology** and **economics**.

SDG 2 Zero Hunger

Hunger is a lack of food. Zero hunger implies that each individual should get enough food for sound health. It is not only the amount of food, but also the quality of food that is important. UNO is working for a hunger-free – world by 2030. In 2020, number of people suffering from hunger were anywhere between 720 million to 811 million. In addition to this, approximately 2. 4 billion people were food-insecure; *i.e.*, they did not have regular access to adequate food. Hunger rose by 161 million in comparison to the previous year. The extent to which the removal of hunger would take place depends on the state of sustainable agriculture in a province of a country. Ensuring the availability of food for a large population of China is a challenge. It calls for smart use of land and sea, technological advances, and sharing of information. Malnutrition and human health are associated with hunger. UN has set the following targets.

Targets (UNDESA, 2023)

2.1 Ending hunger and increasing access to food,

2.2 Ending all forms of malnutrition,

2.3 Agricultural productivity,

2.4 Sustainable food production systems and resilient agricultural practices

2.5 Genetic diversity of seeds, cultivated plants, and farmed and domesticated animals

SDG 3 Good Health and Well-being

Mission Statement

"Ensure healthy lives and promote well–being for all at all ages"

Immunization coverage for the first time in 10 years has dropped. Deaths from TB and malaria have increased. Targets to control have been set in such a way that co-benefits for forests and people are maximized.

Total 13 targets to create action to promote health and well-being for all have been identified. The implementation of SDG 3 depends on many other SDGs. Inconsistencies in SDGs have been pointed out (Mcfarlane, *et al*. 2019) ; *e.g.*, an attempt to increase GDP defined in a conventional way while preserving natural capital (forest reserves) has been made. A short–term gain in human health by forest conservation inflicts direct and indirect health risks for humans as well as other living organisms. Greater exposure to green space, which includes forests, provides mental and physical health benefits for the global urban population which is on the rise. This is shown in Fig. (**1**).

Women are experiencing increased levels of anxiety, depression, and sleep issues than ever before. A recent survey suggested that dissatisfaction with the treatment received by society could be a causative factor (Dowthwaite–Walsh, 2023). A study of psychological well-being (Xi, *et al.*, 2018) concluded that women enjoyed higher levels of purpose in life. The role of altruism in gender differences was examined by these investigators. Women were found to be richer in altruistic behaviors and attitudes.

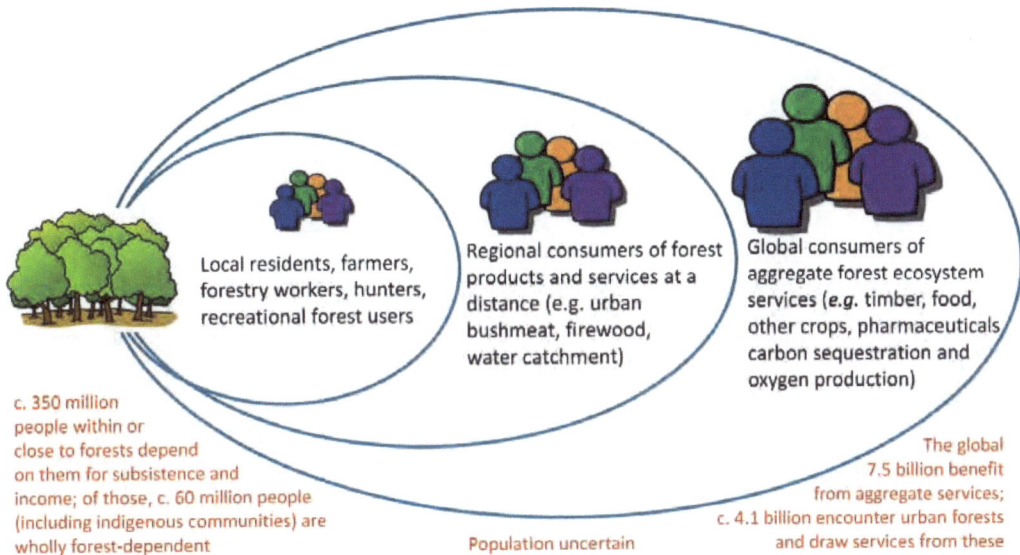

Local residents, farmers, forestry workers, hunters, recreational forest users

Regional consumers of forest products and services at a distance (e.g. urban bushmeat, firewood, water catchment)

Global consumers of aggregate forest ecosystem services (*e.g.* timber, food, other crops, pharmaceuticals, carbon sequestration and oxygen production)

c. 350 million people within or close to forests depend on them for subsistence and income; of those, c. 60 million people (including indigenous communities) are wholly forest-dependent

Population uncertain

The global 7.5 billion benefit from aggregate services; c. 4.1 billion encounter urban forests and draw services from these

Fig. (1). A link between forest conservation and health risks. Source: Mcfarlane, *et al.*, 2019. Good Health and Well being – Framing targets to maximize co-benefits for forests and people, Chapter 3. In: Sustainable development goals: their impacts on forests and people, P. Katila, C. J. Pierce Colfer, W. de Jong, *et al.* (editors) Cambridge University Press, 72 – 107, 2019. Targets (UNDESA, 2023).

Targets (UNDESA, 2023)

All the targets are not mentioned here. Only a few of them that are achievable are mentioned.

3.1 Maternal Mortality

Reduce the global maternal mortality ratio to less than 70% per lakh live births by 2030.

3.2 Neonatal and child mortality

End preventable deaths of newborns and children under 5 years of age, with all countries aiming to reduce neonatal mortality to at least as low as 12 per thousand live births and under-five age mortality to at least as low as 25 per thousand live births.

3.3 End the epidemics of AIDS, tuberculosis, malaria, and neglected tropical diseases, and combat hepatitis, waterborne diseases, and other communicable diseases.

3.7 Sexual and reproductive health

Ensure access to sexual and reproductive health care services for all

3.8 Universal Health Coverage

3.9 Environmental Health

Substantially reduce the number of deaths and illnesses from hazardous chemicals and air, water, and soil pollution and contamination.

Means of implementation for the targets include

3.a Tobacco control

3.b Medicines and vaccines

3.c Health financing and workforce

3.d Emergency preparedness

Strengthen the capacity of all countries, developing countries in particular, early warning, risk reduction, and management of national and global health risks (UNDESA 2023)

SDG 4 Quality Education

Mission Statement

Ensure *Inclusive* and *Equitable* Quality Education and Promote Lifelong Learning Opportunities for all.

The phrase 'Inclusive' means no discrimination based on age, sex, race, ethnic origin, gender, *etc.*, and the phrase 'equitable' means reasonable and fair.

To evolve an actionable plan, the following targets (UNDESA, 2023) were identified.

4. 1 Free primary and secondary education.

4, 2 Equal access to quality pre-primary education.

4. 3 Equal access to affordable technical, vocal, and higher education.

4.4 Increase the number of people with relevant skills for financial success.

4.5 Eliminate all discrimination.

There are three more targets. The chapter skips them with the suggestion that readers should refer to the relevant literature; *e.g.*, UNO official website.

Higher Education Sustainability Initiative was put in place to discuss the transformation of higher education due to the impact of the COVID–19 pandemic. The first webinar was conducted on April 27, 2022. The topic of the webinar was the Transformation of Higher Education post-COVID - 19.

Two related questions are:

1. How can it be decided that the education received by a student is quality education?
2. What is quality education?

Quality education is defined as education that focuses on the emotional, physical, mental, cognitive, and social development of each student without any consideration given to gender, race, ethnicity, socioeconomic status, or geographical location. The main concern is how SDG 4 can be achieved. UNICEF, a UN agency that works in the area of child education, is committed to improving the quality of education given to children. According to UNESCO's global education monitoring report, 2017 – 2018 (UNDESA, 2023), 264 million children and youth were not going to school. Everyone has a vanguard role to play be it students, teachers, parents, schools, and governments.

As far as schools are concerned, the following points are vital (India Today Web desk, October 11, 2018)

1. Ensuring enrollment and attendance.
2. Minimizing dropouts.
3. Prioritizing the concept of ' learning for all'.

It involves receiving instructions from well-trained teachers equipped with resources and well-designed curriculum.

1. Distribution of mid-day meals in schools in rural areas, and
2. Strict implementation of the National Education Policy and frameworks contained in it. This is mandatory for all countries.
3. Caring for adult education, and for colleges of the university.
4. Impart value education to engineering students.

SDG 5 Gender Equality

Mission Statement

Achieve Gender Equality and Empower all Women and Girls

Less than 30% of researchers in science are women. According to the latest data available up to June 2019, pertaining to Research and Experimental development, the ratio of women scientists to the total number of researchers for countries of the world was computed. The regional averages are based on the available data only for staff employed both full-time and part-time. Regional averages for 2016 were computed. These are shown in Fig. (**2**).

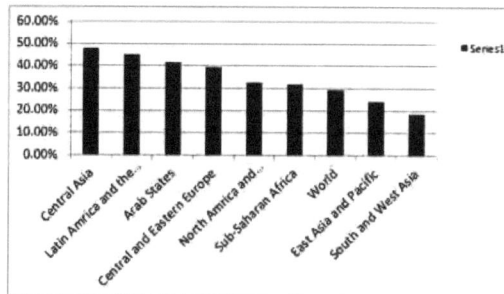

Fig. (2). The ratio of women scientists to the total number of researchers for countries of the world based on full-time equivalents. Head counts, except for Congo, India, and Israel, were used; for China total personnel in R & D activities, were used.

Qualitative Factors

1. Cultural values; *e.g.*, in India, it is believed that women are assumed to be born for household work (cooking food) and reproduction.
2. Challenges of gaining knowledge of difficult subjects (Science, Technology, Engineering and Mathematics).
3. Conflict between male and female members of the family in case a female member is working; who will look after the child.

A little can be done about these qualitative factors affecting the gender gap. Increasing the number of awards for the achievements of women in scientific research would serve as an incentive for a female child to opt for science after High School.

Gender Equality and Women Empowerment are hot subjects in social sciences. Representation of women in workplaces, value chain, and in society needs to be ensured in such a way that no woman is left behind. The UN is not optimistic about achieving this goal by 2030. Nearly half of the married women lack decision-making power over their sexual and reproductive health and rights.

Targets

5.1 Discrimination against all females in all forms is to be ended.

5.2 All forms of violence against all women and girls in the public and private spheres, including trafficking and sexual and other types of exploitation, are to be eliminated.

5.3 Harmful practices, such as child, early and forced marriage, and female genital mutilation, are to be eliminated.

5.4 Unpaid care and domestic work through the provision of public service, infrastructure, and social protection policies and promotion of shared responsibility with the household and the family as nationally appropriate, are to be recognized and valued.

5.5 Women's full and effective participation and equal opportunities for leadership at all levels of decision–making in political, economic, and public life are to be ensured.

5.6 Universal access to sexual and reproductive health rights and agreed in accordance with the program of action of the International Conference on Pollution and Development and Beijing platform for action and the outcome documents of their review conferences is to be ensured.

There are indicators to evaluate how far a given target has been achieved.

SDG 6 Clean Water and Sanitation

These are two basic needs of human life. Population pressure and irresponsible behavior of individuals are two factors responsible for the pollution of water resources. The supply of drinking water to households in any city of the world is a challenging task. Water purification technologies are available. But, the real concern is the cost. A few of them are described below:

Electro-dialysis uses ion exchange membranes both *pure and hybrid*. **Ion Exchange Membranes (IEMs)** generate secondary wastewater while regenerating ion exchange resins. Super- omni-phobic surfaces (return values of contact angles >150 with low contact angle) with hierarchically arranged scales of texture are used in direct contact membrane distillation. **Hybrid Ion exchange membranes** combine ion exchange and pressure-driven membrane processes (microfiltration, ultra-filtration, nano-filtration) together. **Ion concentration polarization** (ICP); a phenomenon of ion transport when ions selectively pass through a membrane occurs through strong capillary actions.

Sanitation is a basic requirement for a civic life. Creation and maintenance of sanitary facilities for families living in a city involve planning and cost. It is the responsibility of municipal corporations. Public health conditions critically depend on the disposal of human waste (excreta and sewage). Poor sanitation may lead to the transmission of diarrhoeal diseases; e. g, cholera, dysentery, and several others. Swachchha Bharat Abhiyan is an initiative of the Prime Minister of India for the rural poor in this direction.

Targets (UNDESA, 2023)

6. 1 Universal and equitable access to safe and affordable drinking water is to be achieved by 2030.

6.2 Access to adequate and equitable sanitation and hygiene for all is to be achieved by 2030.

6.3 Water quality is to be improved by reducing pollution, eliminating dumping and minimizing the release of hazardous chemicals and materials, and by substantially increasing recycling and safe reuse globally by 2030.

6.4 Water use efficiency across all sectors is to be substantially increased and supply of freshwater to address water scarcity and substantially reduce the number of people suffering from water scarcity is to be ensured by 2030.

6.5 Integrated management of wastewater at all levels including through transboundary cooperation is to be achieved by 2030.

6. 6 Water-related ecosystems, including mountains, forests, wetlands, rivers, aquifers, and lakes, are to be protected and restored by 2030

6. a International cooperation and capacity-building support to developing countries in water and sanitation-related activities and programs, including water harvesting, desalination, water efficiency, wastewater treatment, recycling, and reuse technology is to be expanded by 2030.

6. b The participation of local communities in improving water and sanitation management is to be supported and strengthened by 2030.

SDG 7 Affordable and Clean Energy

This comes under the UNO Environment Program.

Increased use of renewable energy resources with less dependence on fossil fuels is advisable.

Clean Energy Sources

1. Wind,
2. Solar,
3. Hydrothermal,
4. Geothermal,
5. Bio-energy, and
6. Nuclear.

These are sources of energy that do not pollute the environment. Solar energy (conversion of photons into electricity) has problems associated with conversion efficiency. In the case of nuclear energy, waste disposal is a challenge. The cost per unit of power from hydrogen fuel cells is greater than that from other sources. This scenario would change as technology advances.

Targets (UNDESA, 2023) linked to the environment are mentioned here.

Target 1 Universal access to affordable, reliable, and modern energy sources is to be ensured by 2030.

Target 2 The share of renewable energy in the global energy mix is to be substantially increased by 2030.

Target 3 Global rate of improvement in energy efficiency is to be doubled by 2030.

Target 7.a International cooperation to facilitate access to clean energy research and technology, including renewable energy, energy efficiency and advanced and cleaner fossil fuel technology is to be enhanced, and investment in energy infrastructure and clean energy technology is to be promoted by 2030.

Target 7.b Expansion of infrastructure and up-gradation of technology for supplying modern and sustainable energy services for all in developing countries, in particular least developed countries, small island developing states, and land-locked developing countries in accordance with their respective programs of support is to be carried out.

SDG 8 Decent Work and Economic Growth

Two basic attributes of decent work are:

1. Realization of the fact that workers of industry have fundamental rights with regard to public health safety and remuneration, and
2. Implementation of environmental safeguards; *e.g.*, processing of industrial waste prior to its dumping in the river,

Elevated levels of lead and cadmium in blood and urine samples of those addicted to marijuana have been found (McGraw, *et al*. 2023).). This is the outcome of a study conducted by the National Health and Nutrition Examination Survey with 7000 participants.

Economic growth is an increase in the production of goods and services in an economy. Median projection in a Reuters survey of economists, India's GDP rose to 7.7 percent in the April – June quarter, and 6.1 percent in the previous quarter of the year 2022. UNO advocates for ***inclusive and sustainable*** economic growth. The creation of conditions conducive for the people to have quality jobs that stimulate the economy while no damage is inflicted on the environment is considered.

COVID–19 has influenced the envisioned path of this SDG; loss of jobs, global recession, *etc*. are its milestones. The pandemic hit the world just after four years of adoption of the Addis Ababa action agenda. Naturally, progress made in implementation till the time the pandemic arrived is not much. It has derailed the path of advancement. International Monetary Fund thinks that a global recession is impending. According to an estimate by ILO, nearly half the global workforce faces the risk of loss of their livelihoods. UNO, Department of Economic and

Social Affairs, has set 10 targets and 2 associated ones. Only the first three are mentioned here.

Targets (UNDESA, 2023)

8.1 Sustenance of per capita economic growth in accordance with national and international circumstances and, in particular, at least 70% gross domestic product growth per annum in the least developed countries is to be ensured.

8.2 Higher levels of economic productivity through diversification, technological upgrading, and innovation, including a focus on high–value added and labor–intensive sectors are to be achieved.

8.3 Development–oriented policies supporting productive activities are to be implemented. Decent job creation, entrepreneurship, creativity, and innovation are to be encouraged.

SDG 9 Industry, Innovation and Infrastructure

A new scientific discovery leads to **innovation**. An industry develops around an innovation after the incubation period. Industry means mass production. Two key challenges associated with mass production are:

1. Management of the workforce
2. Quality control of the output

To set up an industry, basic infrastructural facilities; *e.g.*, road, electricity, and water supply are essential. We need industries that voluntarily honor environmental safeguards; e. g., the Central Pollution Control Board in Delhi looks into the matter of industries on the bank of the river, Yamuna pollutes it. Those industries that demand *incentives and investments* in order to meet the standards set by the state are also welcome.

Targets (UNDESA, 2023)

Target 9.1 The development of quality, reliable, sustainable and resilient infrastructure, including regional and trans-border infrastructure to support economic development and human well–being, with a focus on affordable and equitable access for all is to be guaranteed.

Target 9.2 Promotion of inclusive and sustainable industrialization is to be fostered by 2030. Industry's share of employment and gross domestic product is to be increased and doubled into share in least developed countries.

Target 9.3 Access of small–scale industrial and other enterprises, in particular developing countries, to financial services, including affordable credit, and their integration into value chains and markets, is to be increased.

Industries are needed, but they must follow rules and regulations set by the state with regard to damage to the environment, which is the center of all SDGs. This chapter does not provide an exhaustive list of targets and their indicators; instead, it emphasizes on most material subject matter for a clear idea of how innovation gets incubated in the incubation center and leads eventually to a start–up. India has launched many such start-ups in recent years. Only a few of them would survive the challenges of finance in the second year.

SDG 10 Reduced Inequalities

The first question that comes to mind concerns with the measurement of these inequalities. UNO, Department of Economic and Social Affairs, has identified 14 parameters which are called '**indicators**'. This book does not provide a list of them. The institution has set **10 targets** in order to develop an actionable plan. Readers are referred to the website of the organization for lists of both targets and indicators. Inequalities exist both within and between countries. At the level of an individual, inequalities with respect to:

1. age
2. sex
3. disability
4. race
5. ethnicity
6. origin
7. religion

are found. Following targets (UNDESA, 2023) have been set.

Targets

10.1 Inequalities in income are to be reduced.

10. 2 Universal social, economic, and political inclusion is to be promoted.

10.3 Equal opportunities are to be ensured and discrimination is to be ended.

10.4 Fiscal and social policies that promote equality are to be adopted.

10.5 Regulation of global financial markets and institutions is to be improved.

10.6 Representation of developing countries in financial institutions is to be improved.

10.7 Responsible and well-managed migration policies are to be introduced.

10.8 Special and differential treatment for developing countries are to be ensured.

10.9 Development assistance and investment in least developed countries are to be arranged.

10.a Transaction costs for migrant remittances are to be introduced.

All targets are to be met by 2030.

Equitable resource distribution is the central theme of this goal. The key indicator to evaluate the progress of implementation is given below.

Indicator 10.1.1 Growth rate of **household expenditure or per capita income** among the bottom 40% of the population and the total population in the UNSDG framework.

SDG 11 Sustainable cities and communities

Mission Statement

Make cities inclusive, safe, resilient, and sustainable (UNDESA, 2023) .

More than half of the world's population live in cities. It is projected that 7 out of 10 people will likely live in urban areas. More than 80 per cent of global GDP is contributed by cities; drivers of economic growth. UN has identified eight targets (UNDESA, 2023). Only Target 1 and 4 are mentioned in this chapter.

Target 11.1

Access to adequate, safe, and affordable housing and basic services for all is to be ensured and an upgrade of slums is required.

Target 11.2

Access to a safe, affordable, accessible transport system for all is to be provided. Road safety, notably by expanding public transport, with special reference to those in vulnerable situations, women, children, persons with disabilities and older persons, is to be ensured.

Target 11. 3

Inclusive and sustainable urbanization and capacity for participatory, integrated and sustainable human settlement planning and management in all countries are to be ensured.

All the targets are to be achieved by 2030. For the rest of the targets, readers are referred to UN, Department of Economic and Social Affairs. Two key terms from the above discussion are: 1) **Sustainable Urbanization**, and 2) **Resilient Human Settlement**. The former refers to the balance between urban development and protection to the environment with a commitment to equity in employment, shelter, basic services, social infrastructure, and transformation (Anderberg, 1998). The latter refers to the ability to resist, absorb, accommodate and recover from, adapt to and transform in the face of shock; e. g., incidence of an epidemic or pandemic (COVID – 19).

SDG 12 Responsible Consumption and Production

This goal comes under the Department of Economic and Social Affairs, United Nations Organization.

Responsible consumption is essential for a sustainable society.

Mission Statement

Ensure Sustainable Consumption and Production Patterns (UNDESA, 2023)

Irresponsible utilization of natural resources has led to the present predicament that we are in. UN has identified eleven targets to create action for responsible consumption and production. Only a few find mention in this chapter.

Target 12.1 The 10-year sustainable consumption and production framework is to be implemented.

Target 12. 2 Sustainable management and use of natural resources are to be ensured.

This target is to be achieved by 2030. Reader's attention is now drawn to target 12. 5

Target 12.5 Substantial reduction in waste generation is to be achieved.

By 2030, municipal waste is to be reduced appreciably through prevention, reduction, recycling, and reuse. The next target which finds a mention in this chapter is target 12.

Target 12.8 Market distortions that encourage wasteful consumption are to be removed. Rationalization of inefficient fossil–fuel subsidies that encourage wasteful consumption is to be carried out. (Bhagwati, 1971) .

SDG 13 Climate Action

Greenhouse gases; *e.g.*, carbon dioxide, methane, nitrous oxides, and water vapor, cause global warming. Global temperature has already risen by 1.1 degrees C, greenhouse and other emissions are to be reduced to such an extent that their further rise is slow. The 2030 agenda of sustainable development and its 17 goals set by UNO is a blueprint to the action: save lives and livelihoods from the negative impact of climate change.

Mission Statement

Take Urgent Action to Combat Climate Change and its Impacts.

Targets

13. 1 Resilience and adaptive capacity to climate–related hazards and natural disasters are to be strengthened in all countries.

13.2 Climate change measures are to be transformed into policies and planning.

13.3 Knowledge of and capacity for climate change is to be built.

13.4 The UN framework convention on climate change for mobilizing collectively USD 100 billion annually by 2020 for developing countries is to be implemented and the green climate fund is to be made operational.

13.5 Mechanisms for raising capacity for effective climate change - related planning and management in least-developed countries and small islands are to be promoted.

Readers are referred to Chapter 2: Biodiversity and climate change of the present book. Steps to reduce our own footprint will be the same as the steps needed to eliminate global emissions. In many cases, thinking only in terms of our own current footprint incentivizes some short-term investments, like relying exclusively on buying offsets that won't address the **climate challenge**.

An SDG summit to launch the first global report on Climate and SDG synergies.

SDGs 14 and 15 are in the realm of ecological studies.

SDG 14 Life Below Water

An example is the **Diatom to Whale** food chain in the ocean.

Level 1 Carnivores consumers (Jelly Fish, Small Fish, Crustaceans, Sea Stars)

Level 2 Krill, Shrimp, Surgeon fish, turban, nerik

Level 3 Squid

Level 4 Carnivores that eat other carnivores

Level 5 Albatros, Dolphin, Shark, Whale

Maintenance of *homeostasis;* self-regulation which stabilizes the system in response to external (physiological) conditions; *e.g.*, maintenance of water-saturated conditions within a leaf in a plant in the water body, osmotic potential in salt–water fish and controlled fluctuations of body temperatures of organisms in homeo - therms (Hallam & Levin, 1986), is a challenge faced by all organisms, big or small. A fish of tiny body size eats plants. Fishes of bigger sizes consume those of smaller sizes. Thus, primary production and its consumption link species at different trophic levels, which represent an interaction between consumer-resource and predator-prey communities.

Density of ecological communities varies both in space and time (May, 2001; Rai, 2013). Exogenous factors control the random variability of species abundance. The resulting resource distribution among competing species is complex. Conditions for long-term coexistence are derived from a Gaussian utilization function (MacArthur 1969, 70, 72, May 2001). A standard deviation, w, and a separation, d, characterize the utilization function. The probability of extinction increases with the severity of environmental fluctuations. Competition coefficients have a stochastic interaction with environmental fluctuations. When the smallest Eigen-value of the competition matrix is greater than, $\sigma 2k-$; where the numerator is the variance of the environment, and k - is a measure of **niche overlap**, community persists (May 2001). Pledge to protect ever declining health of the ocean ecosystem due to anthropogenic pollution, overexploitation, and the effects of global climate change, is meaningless as nations and industries find loopholes to circumvent essential guiding principles. United Nations Convention on the Law of the Ocean and the London Convention and Protocol are examples (Richmond & Bruesser, 2023).

United Nations, Department of Economic and Social Affairs has set 10 integrated targets (UNDESA, 2023). Only three are mentioned here.

Target 14.1 Prevention and significant reduction of marine pollution of all kinds is to be achieved by 2025.

Target 14.2 Management and protection of marine and coastal ecosystems to avoid significant adverse impacts is to be affected by strengthening their resilience, and action for their restoration in order to achieve healthy and productive oceans is to be taken.

Targets 14.3 Impacts of ocean acidification are to be minimized and addressed through enhanced scientific cooperation at all levels.

SDG 15 Life on Land

Quality of life is defined by WHO as an individual's perception of his or her position in life with respect to culture and value systems in which they live. Following are parameters that help us measure the 'quality of life'.

1. Goals,
2. Expectations,
3. Standards, and
4. Concerns

Food is the primary requirement. The productivity of agricultural ecosystems depends on pollination. The decline in the density of pollinators (wild bees and hoverflies) in Great Britain (Powney *et al*., 2019) and honeybees and butterflies worldwide (Johnson *et al*., 2022) have been reported. It can have devastating effects on plants. Their role in maintaining plant diversity needs to be investigated. The conservation of pollinator density is mandatory as these pollinators can not be replaced by other species.

The Egyptian vulture is a migratory bird. The population of Egyptian vultures in Europe is on the decline. Loss of habitat, decrease in the food supply, collisions with electricity transmission lines, and poisoning from agrochemicals in both Europe and sub-Saharan Africa are causative factors for population decline. IUCN red list marks its status as an endangered species. Conservation efforts are underway as a realization of scavenger services provided by this avian species (Terraube *et al*. 2022). Scientists have been working on methodologies to conserve biodiversity. Environmental DNA to conserve biodiversity (Cristescu & Hebert, 2018) is a methodology that helps the census of species on a global scale in near real-time. Complexities introduced by its misuse have been explored.

Natural history of eDNA: 1) origins, 2) state, 3) lifetime, and 4) transportation have been developed. This method of conserving biodiversity is limited by uncertainties in data interpretation.

Targets

15.1 Conservation, restoration, and sustainable use of terrestrial and inland freshwater ecosystems are to be ensured along with their services; in particular forests, wetlands, mountains, and dry- lands in line with obligations under international agreements by 2020

15.2 Sustainable management of all types of forests is to be implemented, deforestation is to be halted, and degraded forests are to be restored globally by 2020 by afforestation.

15.3 Desertification is to be combated. Degraded land and soil are to be restored by 2030.

SDG 16 Peace, Justice, and Strong Institutions

Peace

Peace is defined as a situation or a period marked by the absence of war in a country or an area. It is facilitated by the principle of peaceful coexistence. Economic Conflict among countries in the world generally leads to war. It can happen between two neighboring countries or countries in two different alliances, *e.g.*, NATO and the EUROPEAN UNION. NATO is a military alignment on the other hand it is a non–aligned movement raised by Jawaharlal Nehru, 65 Nasser, and Desmond Tito advocated for non–alignment. Its objective is 'peaceful coexistence'. The five guiding principles of non–aligned movement are:

1. To encourage friendly relations among countries,
2. To advocate peaceful settlement of international disputes,
3. To oppose the use of force and the use of nuclear weapons
4. To protect human rights
5. To protect the environment

Justice

According to Plato, justice is balance and harmony (Bhandari, 2022). It refers to conflicting aspects within an individual or a country. There are four kinds of justice: ***Distributive justice, retributive justice, criminal justice, social justice, environmental justice, and global justice***. Utilitarianism is a theory that advocates actions in favor of 'pleasure' and opposes actions that oppose it. ***Distributive justice*** is a term in Social Psychology. It is the perceived fairness of how rewards and costs are shared by group members. A moral theoretic alternative to *utilitarianism* was put forth by John Rawl in his book 'A Theory of Justice' in 1971. A critique of this theory of justice is found in a book by welfare economist, Amartya Sen, published in 2009: 'The Idea of Justice'.

Strong Institutions

Strong institutions are ones that have the following attributes:

1. Promotion of peaceful and strong societies for sustainable development
2. Access to justice for all
3. Effective, accountable, and inclusive institutions at all levels (*e.g.*, family, Village Panchayats; District and Legislative councils at the state level)

Actions to achieve this goal are encrypted in the following targets.

Targets

16. 1 Significant reduction in all forms of violence and related death rates everywhere is to be achieved.

16.2 Abuse, exploitation, trafficking, and all forms of violence against and torture of children are to be ended.

16. 3 The rule of law at the national and international levels is to be promoted and equal access to justice for all is to be ensured.

16.4 Illicit finance and arms flows are to be reduced. The recovery and return of stolen assets are to be ensured and all forms of organized crime are to be combated by 2030.

SDG 17 Partnerships for the Goals

Mission

To Strengthen Means of Implementation and to Revitalize the Global Partnership for Sustainable Development (UNDESA, 2023)

The following definition of sustainable development is given in the Brundtland Report, 1987.

"The development that meets the needs of the present generations without compromising the ability of future generations to meet their own needs"

At Earth Summit, Rio de Janeiro Brazil, more than 178 countries adopted Agenda 21; Global Partnership to improve quality of life which is dependent on how little damage we inflict on the environment. The active participation of Civil Society, the Private Sector, Governments, the UN system, and NGOs to raise funds for the implementation of the Addis Ababa Action Agenda is solicited. Financial assistance to developing countries for the creation of facilities for clean water and sanitation is required. Provision for Debt relief to developing countries by developed ones is recommended. The International Monetary Fund and World Bank are two institutions that facilitate developmental activities in poor countries by providing soft credit and loans. Countries rich and poor can directly participate in such cooperation. In order to generate an action plan to achieve this goal, nineteen targets (UNDESA, 2023)have been identified. These are classified into five categories: finance, technology, capacity building, trade, and systematic issues. Only a select few are mentioned here

17.11 Exports of developing countries, in particular with a view to doubling the least developed countries (LDC) share of global exports are to be increased significantly by 2020

17.18 The global partnership for sustainable development is to be revitalized.

17.19 Existing initiative to develop measurements of progress on sustainable development that complement GDP, and support statistical capacity building in developing countries is to be expanded.

Indicators

1. Total amount of funding for developing countries to promote the development, transfer, dissemination, and diffusion of environmentally sound technologies
2. Number of countries with mechanisms in place to enhance policy coherence of sustainable development.

CONCLUSION

Achieving these goals is a challenge. Cooperation among member nations with minimal conflict and maximum sacrifice on the part of an individual country is the only hope. Following are three grand challenges.

Grand Challenges

1. No poverty,

2. Zero hunger
3. Zero greenhouse emissions

The first two are SDGs 1 and 2. Another relates to the notorious greenhouse effect. Total eradication of poverty (Swindle, 2013) from the world is not possible in the near future. Consider the case of poor countries in Latin America and the Caribbean region; Haiti is an example. It is one of the poorest countries in the world. Its Gross National Income (GNI) per capita was USD 1420 in 2021. Its rank is 163 on a scale which ranges from 1 to 191. How poverty line is marked in India? For urban areas, it is 1,286 INR per month, and for rural areas, it is 1059.42 INR per month. Individuals earning less money than this are poor. The poverty line in the US has nothing to do with that of India. Therefore, there is a need to define absolute universal poverty. Certainly, it would not be linked with Gross Domestic Product (GDP).

Metabolic processes, which regulate physiological parameters within a range that the individual organism can tolerate, lead to a phenomenon known as *homeostasis* (Billman, 2020). It is a challenge to ensure the maintenance of homeostasis in plants; green or otherwise. It could be due to the enhanced level of air pollution or extreme weather conditions. The existence of **Green Space is critically dependent on the maintenance of** *homeostasis* **both on land and at sea. Without it, there will be no agricultural productivity; therefore, no Sustainable Agriculture.** Enhanced demand for food under the current rate of global population increases coupled with increasing energy-related demand, which raises the level of greenhouse gas emissions (UNEP, 2023) and ruptures the agricultural production systems. Harvesting technologies under practice add to this damage. Developing environmental safeguards and putting awareness and education programs in place is the objective of a **World Bank initiative** started in Indian State Universities a few years back. State universities in other countries have also started similar initiatives.

Food grain production to feed everyone in the world is necessary. It is a difficult target to achieve. Adequate measures to control the growing population of a country must be taken; otherwise, it would be a nightmare. Food is the minimum that animals and humans need. Humans require more than that. They require varieties of freedom to grow: *freedom of food choice, freedom of occupation, freedom to follow a religion, etc*. Latter refers to a democratic society. The reader is challenged with the following: Freedom is both the end and means of sustainable economic life and holds the key to secure the general "welfare "of the world's population. Is this idea realizable?

ACKNOWLEDGEMENT

Sarita Kumari is thanked for helpful discussion.

REFERENCES

Arthur, R.M. (1969). Species, Packing or what competition minimizes. *Proc. Natl. Acad. Sci. USA, 64*(4), 1369-1371.
[http://dx.doi.org/10.1073/pnas.64.4.1369]

Anderberg, S. (1998). Industrial Metabolism and the linkages between economics, ethics and the environment *Ecol. Econ, 24*(2 – 3), 311.

Bhandari, D.R. (2022). Plato's Concept of Justice: An Analysis *20th World Conference on Peace.*Boston

Billman, G. E. (2020). Homeostasis: The underappreciated and far too often ignored central organizing principle of physiology *Plant Physiol 11.*
[http://dx.doi.org/10.3389/fphys.2020.00200]

Bhagwati, J.N. (1971). The generalized theory of distortions and welfare.*Trade, Balance of Payments and Growth in honor of Charles P..* Amsterdam: North – Holland.

Cristescu, M.E., Herbert, P.D.N. (2018). Uses and misuses of environmental DNA in Biodiversity Science and Conservation. *Annu. Rev. Ecol. Syst., 49*(1)
[http://dx.doi.org/10.1146/annrev – ecolsyste- 110617 – 062306]

Hallam, T.G. (editors)*Mathematical Ecology: an Introduction.* 196, Springer - Verlag: Berlin..

India Today web-desk, New Delhi, October 11, 2018..

Kumari, S., Upadhyay, R.K., Kumar, P., Rai, V. (2021). Dynamics and patterns of species abundance in ocean: A mathematical modeling study. *Nonlinear Anal. Real World Appl., 60*, 103303.
[http://dx.doi.org/10.1016/j.nonrwa.2021.103303]

L. Dowthwaite – Walsh. (2018). The Paradox of Progress: Why More Freedom is not making women happier. *Neurosci. News.*

Lorenz, E.N. (1963). Deterministic Non – periodic flow. *J. Atmos. Sci., 20*(2), 130-141.
[http://dx.doi.org/10.1175/1520-0469(1963)020<0130:DNF>2.0.CO;2]

Leibenstein, H. (1974). An interpretation of the economic theory of fertility: promising path or blind alley? *J. Econ. Lit., 12*, 457-479.

McGraw, K.E., Nigra, A.E., Klett, J., Sobel, M., Oelsner, E.C., Navas-Acien, A., Hu, X., Sanchez, T.R. (2023). Blood and urinary metal levels among massive marijuana users in NHANES (2005 – 2018). *Environ. Health Perspect., 131*(8), 087019.
[http://dx.doi.org/10.1289/EHP12074]

Madhav, N., Oppenheim, B., Gallian, M. (2017). Pandemics: Risks, Impacts, and Migration.*Disease Control Principles: Improving Health and Rducing Poverty.* The International Bank for Reconstruction and Development /The Work Bank.

MacArthur, R. (1970). Species packing and competitive equilibrium for many species. *Theor. Popul. Biol., 1*(1), 1-11.
[http://dx.doi.org/10.1016/0040-5809(70)90039-0]

MacArthur, R.H. (1972). *Geographic Ecology..* New York: Harper & Row.

May, R.M. (2001). *Stability and Complexity in Model Ecosystems..* Princeton, New Jersey: Princeton University Press.
[http://dx.doi.org/10.1515/9780691206912]

Ngonghala, C.N., Plucinski, M.M., Murray, M.B. (2014). Poverty, Disease and the Ecology of

ComplexSystems. *PLoS Biol.,* (April),
[http://dx.doi.org/10.1371/journal.pbiol.1001827]

Nelson, R.R. (1956). A Theory of the low - level equilibrium trap in the underdeveloped economies. *Am. Econ. Rev., 46*(5), 894-908.

Our world in Data team, Reduce inequality within and among countries, published online at OurWorldInData.org.

Powney, G.D., Carvell, C., Edwards, M., Morris, R.K.A., Roy, H.E., Woodcock, B.A., Isaac, N.J.B. (2019). Widespread losses of pollinating insects in Britain. *Nat. Commun., 10*(1), 1018.
[http://dx.doi.org/10.1038/s41467-019-08974-9]

Rawl, J.A. (1971). *A Theory of Justice..* Penguin Books.
[http://dx.doi.org/10.4159/9780674042605]

R. A. Mcfarlane, J. Barry, gueladio Cisse, et al. SDG 3; Good Health and Well being – Framing targets to maximize co benefits for forests and people, Sustainable Development Goals: Their Impacts on Forests and People, P. Katila, C. J. Pierce Colfer, W. de Jong ,et al. (editors) Cambridge University Press, 72 – 107, 2019.

Rai, V. (2013). *Spatial Ecology: Patterns and Processes.* Bentham Science Publishers: Sharjah, UAE.
[http://dx.doi.org/10.2174/97816080549091130101]

Richmond, R., Buesseler, K. (2023). The future of ocean health. *Science, 381*(6661), 927.
[http://dx.doi.org/10.1126/science.adk5309]

Sen, A. (1997). *Development as Freedom..* Oxford University Press.

Sen, A. (2009). *The Idea of Justice..* Penguin Books.

Salmun, H., Molod, A. (2006). Progress in modeling the impact of land cover change on the global climate. *Prog. Phys. Geogr., 30*(6), 737-749.
[http://dx.doi.org/10.1177/0309133306071956]

Swindle, J. (2013). Poor Economics: A Radical Rethinking about the Nature of Poverty. *J. Human Dev. Capabil., 14*(2), 312-314.
[http://dx.doi.org/10.1080/19452829.2013.785225]

Third Global Conference on Financing for Development. UNO, New York.

Third Global Conference on Strengthening synergies between the Paris Agreement and 2030 Agendafor Sustainable Development, UNDESA.

UNEP. (2023). *Emissions Gap Report.*https//sdgs.un.org/goals

WHOQOL: Measuring Quality of life, WHO archives, May 15, 2020.

Xi, J., Lee, M.T., Carter, J.R., Delgado, D. (2022). Gender differences in purpose in life: The Median Effect of Altruism. *J. Humanist. Psychol., 62*(3), 352-376.
[http://dx.doi.org/10.1177/0022167818777658]

Yaffee, R.A. (2009). An Introduction to State Space Models.

SUBJECT INDEX

A

B

C

W

X

www.ingramcontent.com/pod-product-compliance
Lightning Source LLC
Chambersburg PA
CBHW041449210326
41599CB00004B/185